0053605

DEHPOLO99

F
370
D38
1982

DeHart, Jess.
Plantations of
   Louisiana

$12.95

© THE BAKER & TAYLOR CO.

PLANTATIONS OF LOUISIANA

Plantations of Louisiana

By JESS DeHART
With Melanie Hamlet DeHart

PELICAN PUBLISHING COMPANY
GRETNA 1989

First printing, 1982
Second printing, 1989

Library of Congress Cataloging in Publication Data

DeHart, Jess.
    Plantations of Louisiana.

    Includes index.
    1. Plantations--Louisiana--Guide-books. 2. Historic
buildings--Louisiana--Guide-books. 3. Architecture,
Colonial--Louisiana--Guide-books. 4. Architecture,
Modern--19th century--Louisiana--Guide-books. 5. Louisi-
ana--Description and travel--1981-   --Guide-books.
I. Title.
F370.D38   917.63'0463   82-325
ISBN 0-88289-338-6   AACR2

Manufactured in the United States of America

Published by Pelican Publishing Company, Inc.
1101 Monroe Street, Gretna, Louisiana 70053

CONTENTS

LIST OF MAPS

PLANTATIONS OF LOUISIANA

INTRODUCTION

Home Place (Attakapas)

Introduction _____

This book has been designed to assist readers on nostalgic trips into the interesting and historic past of the fabulous plantation era in Louisiana. Those trips could be weekend jaunts of a spontaneous nature, or perhaps completely planned vacations encompassing several areas that would be self-guided with the aid of the maps within. And, too, there are locations readily reached from any part of Louisiana by just a comfortable drive, regardless of where your travels in the state may take you. Many of these old homes are open to the public for a modest fee; they are noted within.

Included are thirteen maps locating over 300 existing early Louisiana plantation homes. These have been accurately located with the odometer from easily recognized, permanent landmarks, leaving no doubt or guesswork in locating them. It is the most accurate and complete guide of its kind. Also included are examples of architecture that will help you to readily recognize and fully appreciate each one for its own merits as you come upon it.

As compared to ancient European structures or to historic monuments of the other continents, these Louisiana plantation manors are not considered old. However, this is Louisiana's heritage and history, and these 150- to 200-year-old dwellings represent that proud history. Louisiana planters played a major role in the formation of our nation, and these homes are mementos to them. Louisianaians are proud of them.

You can enjoy these gems of past history—not just today, not just next week, but as often as you wish when traveling the historically rich highways of Louisiana.

LOOKING FORWARD TO THE PAST

Looking forward to the Past

Louisiana has a strange and weird beauty that can never be duplicated anywhere else on this earth. Its magical wilderness of a century or two past could just as well be tomorrow. The stage is a setting of great oaks draped with cloaks of grey moss, eerie prehistoric cypresses sparsely topped with furry greens, and masses of willows dripping leaves into waters below that intermingle in misdirection. Lazy bayous seem untouched by man, as if time had stood still since the making of the earth. The jet black soils beneath motionless waters form gigantic mirrors that reflect multitudes of mystical beauties.

Tinted with earth colors, the waters of the great Mississippi wind majestically in an indecisive manner through endless shoulders of green levees and luxuriant farmlands. Opulent soils of grey, black, and red cover every inch with wealth; the river is enriched with such an abundance of nourishment that it overflows onto neighboring lands, bringing lush growth to their adjacent borders.

The mystery of time plays havoc with this world, giving the time no place to go, and lets it hover to create a peace never experienced before. The past will never be fully dead, because there will always be curiosities about it that will not allow it to die. It has a magnetism that draws us to probing into it with a nostalgic tranquility that we all seem to be forever seeking.

If we were each allowed time machines, with dozens of buttons, to select our own era over past centuries, it would be safe to say that most of us would choose the era of the great plantations. History will forbid us to search for a peace in this present age, and we know it does not exist now and will not exist in the future. For this reason, our nostalgic heartstrings return us to a time of gracious leisure and peaceful living, purely represented during the early nineteenth century. This nostalgia can haunt us and possess us without our awareness. And a magical aura surrounds us when venturing into that age, enhanced by the mystery and intrigue of old plantation homes.

To experience this we must travel the lands of old plantation life, void of modern influence. Do not measure the hours with a clock, but allow time to be of no concern. Leisurely place yourselves

17

in the days of the great planters. Picnic in areas undisturbed by man or by the years. Climb atop the levees for panoramic visions of the great Mississippi, which floated elegant paddle boats of the time. Envision the brown waters busy with barges of sugar, tobacco, indigo, and cotton. Listen for the whistle of the steamer docking just below with fancy provisions from afar.

Walk into the fields of work and among the whitewashed Negro cabins. Put your mind away from modern man and machines and let the sounds of plantation life enter your senses so that you can experience a joy and tranquility felt no other place on earth.

Walk the long lanes lined with oaks and magnolias, scented in beauty. Climb the stairs to the great manors, beckoning all friendships long ago lost. Walk into the rooms and up the stairs to the capping belvederes where one has endless views in every direction: off into the fields and mills of work and over vast waterways of travel and wide avenues that reach to the outer world.

Envision neighboring manors with great pillars of white, pink, and green. To the rear from the sugar mill stack rises a mass of vapors that drift lazily into the blue, mingling with snowy clouds that hover to protect workers from a bright and offensive sun, while their glistening knife blades signal that it is a time of harvest. Multitudes of mule-drawn wagons cover field roads with loads of sweetened cane headed for the mill. This, the happy season, will soon produce well because of the hearty efforts of all working together. From the field worker to the hunter, from the household servant to the blacksmith, from the plantation owner to the oversee—all will reap rewards from the fruitful crop that will bear many tons of snow-white crystals of sugar for the tables of many nations.

Place your body and your mind into the aged hands of time, a time when all this was opulence and reality, ushering in fragrances of magnolias, wisterias, and jasmines, filling the breezes, and the sounds of human existence in gaiety and laughter and songs of Negros at work. And blood will once again begin to flow through the veins of bygone life to bring us back again into its own time.

The swishing of invisible silk skirts upon polished stairs will delight our fancy. And a magical old home with bare walls and empty rooms will again come alive, revealing cheerful atmospheres of a colorful, gay, and exciting era, with heavily scented flowers bordering lacy iron-trimmed galleries. This grandeur in decay will once again become rejuvenated.

Allow yourself to listen to the sounds of that past and you too can become part of it. In the parlor with visiting friends, you are free of the annoyance of the electronic devices that can spoil such pleasant times; truly enjoy this visit with old mellow friends over a cheerful glass of brandy or wine. Hear the faint melodious sounds that come from the music room, where lessons on the harp are given. And in the distance the cheerful innocent laughter of children at play can be heard.

Return to the green lawns below and meander toward the vapors of delicious foods of Creole cooking in the outdoor kitchen. Hundreds of preserves in jars stand orderly on shelves of outer pantries beneath the cool refreshing rainwater cisterns. With them are butters, cheeses, and smoked hams neatly stored for future use. Beyond the smokehouse are gardens of colorful vegetables and herbs fenced carefully in protection. Great orchards of fruit and nut lie before the domestic fowl yards and the distant pastures of cows, horses, and sheep.

Pluck a plum, a pear, or some figs from life-giving trees and stroll over to the little white chapel with the gleaming steeple, where one can be spiritually enhanced by a private visit with God. He has given us this peace; why then should we not ask Him to share it with us?

A thoroughbred gracefully grazes the green lawns, immaculately manicuring them. Stables and barns are orderly. Weathered, unpainted cypress fences belong in the fields, looking airy and unrestricting. The atmosphere is filled with songs of birds and cooings from the nearby *pigeon-niers*. Listening to the frivolous giggles of play coming from the garçonnière, should one hazard a guess? With him is that free loving and indecorous belle from the neighboring plantation; papa had better not hear them!

A young Creole maiden in black riding habit, with trailing skirt daringly pinned at her waist, lends a look to her shapely leg. Her hair flows wildly behind; her gestures are graceful and free.

Go past the blacksmith shop and the carpenter shop, meander down to the general store where goods of need are stocked. So many items of nostalgic quality bring back thoughts of the more simple and gracious life. Wait for the *lagniappe* (something extra) before you leave!

Young children are innocently skinny-dipping in the pond beyond the guest house. "Lawdee, I's goona blister dem chilluns behinds!" says tante Marie with a pearly grin. But she hesitates, pretending not to know what they are doing; they are having so much harmless fun!

We know that there will be tragedies, sorrows, and hardships along with the tranquility and gaiety, yet we choose to remain. The conveniences of the modern world are not ever missed if one does not know of them. Our pleasures and happiness are of our own creation and we cherish every moment of them. They are not those made by strangers to be enjoyed among strangers. They are fashioned by friendships and enjoyed in the intimacy of friends where we can take time enough to season lasting relationships.

Hear the gay laughter and song of the Negroes, the ringing of the silver-toned plantation bells, and the whistles from the boat on the river. All create a sense of awareness felt no other time, no other place. Hard work does not destroy this tranquility and peace of mind; on the contrary, healthy attitudes of people together at work and in leisure prompt a more stable state of well-being and happiness. Personal relationships between workers and owners are so much better.

The grandeur of many great manors has been taken by the elements and by man; such monuments as Woodlawn, Belle Grove, and Uncle Sam have been destroyed by fires and the ravages of the Mississippi. Yet there are those that forever stand proud, seeming to endlessly challange all facets of destruction—Parlange, Melrose, Oak Alley, Evergreen, and many others. These beautiful gems seem to always maintain a character that almost approaches human.

Become an active dreamer and join the thousands who have returned to walk the very same pathways under the very same spacious skies and umbrellas of vast oak trees that existed then for those Creole ancestors. They would certainly be pleased to share them with us. Climb the unchanged stairways to their spacious galleries that overlook their timeless waterways and their endless fields.

Join others who have unchained themselves from the binding shackles of demand and for a while break away from stifling powers of responsibility. Take the time to breathe again: breathe the refreshing airs of the past and take time to enjoy the wide open spaces and genial happenings. Imagine occasions of festive costume balls that lasted throughout the night and gala parties with dozens of guests from afar. See the mantels and stairways garlanded with roses and magnolias and the beautifully gowned ladies gracing the ballroom floor with a regal elegance, their gentlemen escorts so greatly pleased. And envision gay and cheerful Christmas time with the Negro workers, or joining together on the river banks with pleasures from the never-to-be-forgotten showboat, with its minstrels, elaborate stage settings, and dazzling costumes. Visit great feasts of unimaginable dishes and merrily decorated tables loaded with numerous varieties of meats, vegetables, fruits, pastries, desserts, and wines. It can be so very rewarding; it will be so very peaceful. Return with us again, time after time, to help dismiss the everpresent pressures created by the thankless and the detached.

The great plantation era was no myth, but was real. It will not die with corroding walls and decaying pillars. There will always be reminders of its existence, even if only through the majestic oaks and crumbling sunbaked bricks. There will always be romanticists who will successfully stir curiosities in our minds and in the warmth of our hearts, exciting our passions for mystery and intrigue. We do not need to be romanticists to enjoy the many pleasures and excitements created by them; by allowing ourselves to be removed from the present we may wander into that never-to-be-forgotten and bewitching past.

ARCHITECTURE

Early Plantations

Early Plantations

It would be very difficult to describe all of the numerous architectural expressions that presently exist in Louisiana, illustrating every past period of fashion and culture. Even the designs within the era of great plantation society seem too numerous to describe. But the majority of structures built in that time did follow certain standards in design that can be categorized by a few styles.

The great plantation era, which covered the period from about 1780 until 1860, saw many changes in lifestyle as well as architecture. From the smaller planter's dwelling to the mansions of vast empires, the architecture of houses projected the qualities and personalities of the individuals who lived in them.

In the beginning, designs were chiefly controlled by needs and comfort, with a direct relation to financial limitations as well as specific needs with regard to the size of the family. Heat, humidity, rainy seasons, the threat of floodwaters, and isolation from cities were major factors that dictated the design of dwellings on the plantation. Isolation from cities brought on the need for additional storage facilities for large amounts of provisions that could not be obtained on the plantation. Provisions including clothing fabrics, and certain foods, implements, and supplies required to sustain life in these frontierlike regions had to last until the next shipments from distant cities arrived.

Steep roofs, high ceilings, large rooms, tall and wide doorways and windows, broad galleries, and thickly walled brick basements were all incorporated to help overcome discomforts and hardships created by nature. Perhaps most importantly, different cultures greatly controlled architectural designs.

CONSTRUCTION

Early Plantation Homes

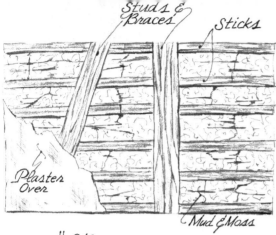

Studs & Braces

Sticks

Plaster Over

Mud & Moss

"*Bousillage*"

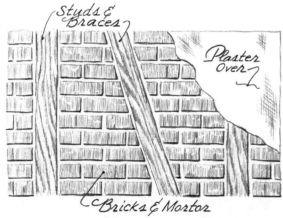

Studs & Braces

Plaster Over

Bricks & Mortar

"*Briquette entre poteaux*"

WALLS

(Paling)

Post & Rail

Snake Rail

EARLY FENCES

Early Plantation Homes

The wall construction style known as *Bousillage* was formed by riverbank clay mixed with moss (or deer hair) molded around sticks and placed solidly between studs. The entire surfaces were either covered with cypress weather boards or plaster, which was made from powdered shells and clay. This structural feature offered very good insulation.

Briquette entre poteaux featured handmade bricks mortared between posts and either plastered over or covered with cypress weather boards. The inside walls were either painted or covered with wall paper. This structure gave good insulation against sounds as well as against the elements.

Windows and doors were large, double, and featured storm protecting shutters with adjustable louvers. Glass panes (when used) were usually somewhat flawed compared to those of modern day, yet they offered a distinctive personality of their own. A special type of window, the dormer, was built out of the roof to allow light into attic rooms.

The roof was covered with either split cypress shingles or rounded or flat tiles. Earlier homes used split cypress shingles; many of those are still in service today, offering a very special aged character that cannot be matched by today's modern materials. Occasionally the roof featured a *belvedere*, an observation gallery with hand railings that was built at the highest point of the roof. The belvedere usually had a small shelter room or cupola at its center point.

Columns surrounding the galleries were either round or square and were fabricated from handmade bricks covered with plaster or were handhewn from cypress. Both types were very durable under all weather conditions.

Gutters were square in design and constructed from milled cypress boards. Cisterns, to ensure an ample fresh water supply, were made with long-lasting cypress staves and were placed several feet above the ground on top of brick storage areas that were used for preserving foodstuffs for long periods. Walkways were covered with either handmade bricks or cypress planks raised slightly above the ground.

Construction products were usually made from easily available local materials and in many cases were fabricated on the construction site. Structural beams and siding and split shingle roofing was made from the abundant cypress; sunbaked bricks from local clay; and plaster and mortar from clam shells and clay. Hardware, hinges, latches, and doorknobs were frequently forged in the plantation blacksmith shop.

Although it was economical and offered very good protection against the weather during all seasons, brick construction for Negro quarters was seldom seen. Most Negroes objected to it, claiming that it brought on rheumatism and imprisoned evil spirits within the house.

There were three basic types of fences used on early Louisiana plantations. The paling fence was used by the French to keep chickens, hogs, and other marauders out of vegetable, fruit, and flower gardens. It consisted of cypress boards driven side by side into the ground with a supporting rail at the top just below the tapered ends. For the post and rail fence, heavy cypress posts were put into the ground and slots were cut through to receive cypress rails. This type of fence was used to confine large stock into certain areas and away from crops. Snake rail fences were introduced into Louisiana by Anglo-Saxons but were traditionally brought to America from Europe by the Scotch-Irish. One advantage of this fence is that it could be easily relocated. However, it required much more material to construct and also took up more space than other fences.

ARCHITECTURAL DESIGNS
Louisiana Colonial

Louisiana Colonial

The Louisiana Colonial style was perhaps the greatest architectural expression developed in the central United States. It began to evolve as early as 1726 with a design that incorporated many features of the West Indies plantation home. At the same time, methods used by Louisiana Indians were very effectively adapted: the bousillage, split cypress shingle, and raised levels.

This design helped to successfully combat the oppressive elements of nature. Variations of the Louisiana colonial can be seen throughout the state, but all were built with common objectives—practical sturdiness, economy, a maximum of comfort, and proper use of space.

Some had galleries in the front and back while on others a gallery completely encircled the dwelling. The high pitched roof, either gabled or hipped, served well to insulate the main level while it offered additional space for loom rooms and *garçonnières* (rooms for the bachelors). Raised high above the ground by brick pillars, the space below the house was walled to form a basement, which was divided into pantries, servant quarters, weaving rooms, wine cellars, carriage rooms, or kitchens. When flooding conditions threatened the area much of the furnishings and supplies were moved to higher levels. Floodwaters did little harm to the brick wall construction, and when it did these damages were easily repaired.

Stairways rose to the floor level of the main house in central areas within the bounds of the gallery, making the basement easily accessible from the main level even under adverse weather conditions. Smaller colonnades of cypress, often beautifully handhewn by artisans, were spaced at intervals surrounding the galleries. These spacious galleries served greatly in several ways—when built very wide, they cooled the summer air considerably before it entered the house, and they offered additional space for living during the hot humid seasons. Most summer days were spent on these galleries, either at work or in relaxation.

The large full length windows and doors opened directly onto the galleries, offering each room individual cooling and access. During stormy or unpleasant weather the louvered shutters were closed for additional protection and insulation.

The Louisiana colonial design seldom included inner or central hallways, since all space was economically utilized. Most Louisiana colonial homes were designed with intimate spaciousness which reflected the warm atmosphere that was familiar to the Creole families housed within them. The entire structure on an average was two and one-half stories high.

One of the best examples of this design is *Parlange* plantation home, below New Roads on False River.

Creole Cottage

J DeHart

Creole Cottage

Usually built by economy-conscious planters of lesser financial means, the early Creole or Creole raised cottage was constructed on brick piers three or four feet above the ground to protect against the ever-present threat of flood. As the conditions against flooding improved with the draining of lands and the building of protective levee systems, the Creole cottage was placed on lower foundations to save on materials and labor.

The average Creole cottage featured a high gabled roof with one or more dormers for the attic rooms. Wide galleries were built across the front and rear of the house with either round, hand-carved columns or box columns with capitals and bases. Wide steps rose to the center of each gallery with banisters and railings bordering the open edges. Attic rooms were reached by stairs at the rear, from either the gallery, kitchen, or dining room.

As conditions demanded, many other features could be added: outdoor kitchens to make additional room in the main house, dovecotes, larger or supplemental cisterns, or additional storage areas.

This design is exemplified by the *White* plantation home on Bayou Lafourche, north of Thibodaux.

Creole Classic

Creole Classic

The Louisiana Creole classic design evolved from the Louisiana colonial with the aim of increasing an effect of grandeur. The floor plan is basically very much the same as that used in the Louisiana colonial, but in the classic style are featured more expressive stairways and columns, and outbuildings usually in the form of wings to the main structure. Belvederes were often added to the roof for panoramic views of the broad fields and the traffic on the waterways.

Galleries were still broad, to serve as useful air conditioners and living areas, but large round columns were added to reach from the ground to the roof, spanning two and sometimes three floors. Inner hallways were incorporated in some new structures as utilization of space became less important. The overall effect remained warm and very attractive, although a good deal more formal than the Louisiana colonial.

One of the best examples of the Louisiana classic style is the popular *Destrehan* plantation home on the River Road near the town of Destrehan.

Greek Revival

J DeHart

Greek Revival

Greek Revival homes were designed for colossal effects. This architecture made its way to the deep South with Anglo-Americans, who in many cases seemed to seek out grand symbols as an effort to elevate themselves socially. By 1840 the Greek Revival influence became quite pronounced not only on plantations but in towns as well.

Although this construction is massive and visually impressive, it reflects a certain coldness that is usually associated with civic landmarks such as city halls and court buildings. Little thought was given by the Greek Revival approach toward practical needs and use of space. It presented quite a formal appearance in contrast to the warm, inviting atmosphere of the earlier plantation homes.

The use of broad galleries was just about entirely ignored, despite their importance in the humid climate. Huge oversize columns seemed to serve no other purpose other than to support the narrow balconies above the entrance doors. Larger than necessary indoor stairways noticeably overpowered the large amount of wasted space they occupied in the perhaps overspacious hallways.

Two and three separate parlors became popular, even if they were not actually needed, while individual family bedrooms and private service rooms seemed inadequate, suffering desperately from lack of sufficient space. Drawing and dining rooms and huge ballrooms, which were usually included, seem far more appropriate for public buildings, where the entire community needs might be in consideration.

There were those who did utilize this design in better fashion, however, by incorporating many practical features of the colonial design. All in all, emphasis seemed to lean toward overpoweringly spectacular effects for the outsider rather than the comforts of those dwelling within the home.

Louisiana Greek Revival design is represented by the imposing *Richmond* plantation home in East Feliciana Parish, east of Norwood, Louisiana.

Other Designs

Other Designs

Various other architectural stylisms were given to a number of plantation homes throughout the state. Many of the designs indicated pompness, others poor taste, while some showed a definite lack of imagination. Others were very interesting and attractive, reflecting pleasing Gothic and Victorian lines. Perhaps the one style that projected the truest example of folly and frivolity in architecture during the entire Louisiana plantation era was the creation of the Steamboat Gothic of the *San Francisco* plantation home. Completely renovated and beautifully furnished, however, it stands today as a showpiece, distinctive and separate from any other home. It stands alone and unique; as an example of an imaginative, yet somewhat strange, architectural development.

MAPS AND DIRECTIONS
Traveling the Plantation Country

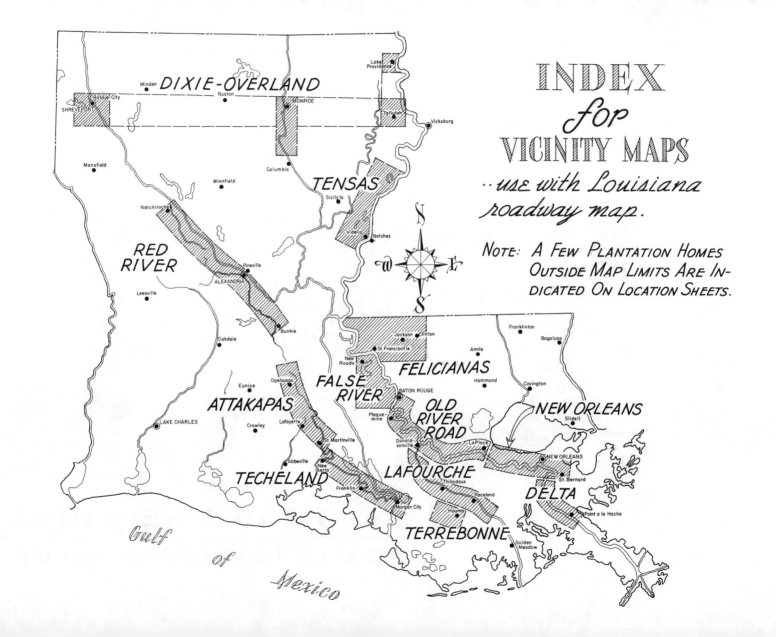

INDEX *for* VICINITY MAPS

..use with Louisiana roadway map.

NOTE: A FEW PLANTATION HOMES OUTSIDE MAP LIMITS ARE IN-DICATED ON LOCATION SHEETS.

DIXIE-OVERLAND

TENSAS

RED RIVER

FELICIANAS

FALSE RIVER

ATTAKAPAS

OLD RIVER ROAD

NEW ORLEANS

TECHELAND

LAFOURCHE

DELTA

TERREBONNE

Traveling the Plantation Country

Most of the plantation homes mentioned in this guide represent a great part of the pioneering past of Louisiana. Perhaps to many viewers some of them may appear small in structure and quite insignificant when compared to the stateliness and grandeur of others. But many of the less imposing homes have been beautifully restored while others have been left to deteriorate very badly. Nevertheless all of them are symbols of an era of great opulence. The planters who initially built each of them have contributed greatly toward the very unique history of the proud State of Louisiana.

When visiting plantation country, the traveler will note that there are many additional reminders of the pioneers who have contributed greatly toward the shaping of Louisiana. There are trappers, fishermen, boatbuilders, Acadian rice farmers and cattlemen, Creole vegetable farmers, and many other practitioners of old skills. Some of the methods of operation and types of living structures, especially the Acadian dwellings, remain today just as they were in the early days.

There are also many old churches, cemeteries, and convents, just as there are numerous slave cabins, sugar mills, and other old plantation buildings that have survived the passing of disturbing years. Spotted throughout the state are museums with hundreds of items of that past: farm implements, household utensils and furnishings, tools, vehicles, clothing, and countless other items to view that can help to better envision the life-styles of those early times. All this adds greatly to the interest, excitement, and enjoyment of traveling through Louisiana plantation lands.

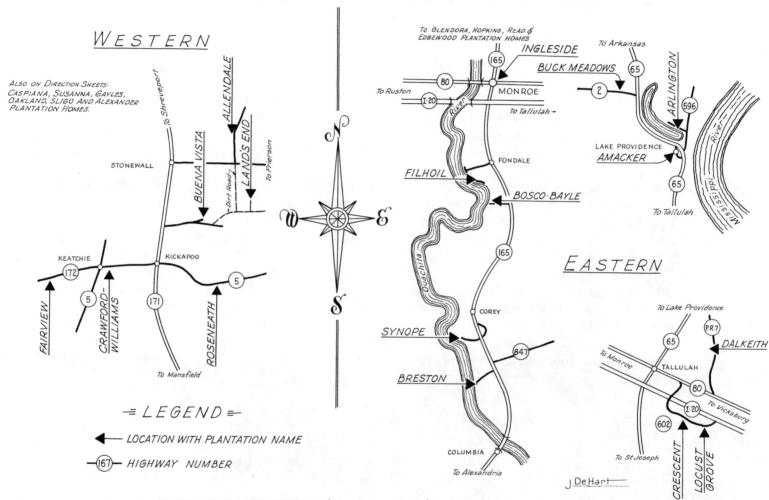

WESTERN

Also on Direction Sheets: Caspiana, Susanna, Gayles, Oakland, Sligo and Alexander Plantation Homes.

ALLENDALE
LANDS END
To Frierson
To Shreveport
BUENA VISTA
STONEWALL
Dirt Road

KEATCHIE
172
FAIRVIEW
5
CRAWFORD-WILLIAMS
KICKAPOO
171
5
ROSENEATH
To Mansfield

N
W
E
S

LEGEND

← LOCATION WITH PLANTATION NAME
167 HIGHWAY NUMBER

To Glendora, Hopkins, Read & Edgewood Plantation Homes
165 INGLESIDE
BUCK MEADOWS
To Arkansas
65
ARLINGTON
80
To Ruston
MONROE
I-20
River
To Tallulah
2
596
LAKE PROVIDENCE
AMACKER
Mississippi River
FONDALE
FILHOIL
BOSCO-BAYLE
65
To Tallulah
Ouachita
165

EASTERN

COREY
847
SYNOPE
BRESTON

To Lake Providence
P.R.7
DALKEITH
65
To Monroe
TALLULAH
80
To Vicksburg
I-20
602
CRESCENT
LOCUST GROVE
To St Joseph

COLUMBIA
To Alexandria

J DeHart

• DIXIE OVERLAND PLANTATION HOMES •

DIXIE–OVERLAND PLANTATION HOMES

SHREVEPORT

At Keatchie, on Hwy 5, 0.2 mile E of intersection with Hwy 172 to *Crawford-Williams Plantation Home* (1848). Modified Louisiana classic, fair condition; private. In the community of Keatchie there are several other old homes and buildings of interest.

From the intersection of Hwy 172 and Hwy 5 at Keatchie, travel 3.8 miles W to *Fairview Plantation Home* (1840). Creole raised cottage; splendid porch with French doors. Family cemetery at entrance, good condition; private.

From the junction of Hwy 171 and Hwy 5 at Kickapoo, travel 4.5 miles E on Hwy 5 to *Roseneath Plantation Home** (1840). Louisiana classic, good condition, beautiful grounds; private.

From the junction of Hwy 5 and Hwy 171 at Kickapoo, travel N 2.4 miles on Hwy 171. Turn right on Red Bluff Rd. and travel E 2.0 miles to entrance road (L), then 0.4 mile on entrance road to *Buena Vista Plantation Home* (1854). Louisiana classic, beautiful grounds; private. Cannot be seen from highway. Entrance road private—do not enter without prior permission.

Travel on Frierson Rd. from the junction with Hwy 171 at Stonewall E 3.7 miles to Linwood Rd. Turn right on gravel road and travel 2.9 miles to the junction with the dirt road. Turn left and travel 0.4 mile to entrance road (L) of *Land's End Plantation Home** (1857). Modified Louisiana classic, good condition; private. The entrance road to Land's End is on private property and should not be entered without prior permission. Inquire at Shreveport Tourist Commission.

Return to the junction of Frierson and Linwood roads. Travel N 0.4 mile on Linwood Rd. to entrance road (L) of *Allendale Plantation Home** (1854). Large log cabin, cannot be seen from road; private.

On the campus of LSU Shreveport on Hwy 1, to the left and rear of the main administration building is *Caspiana Plantation Home* (1852). Louisiana colonial, good condition; open to public. This beautifully renovated home was moved from Caspiana plantation and now serves as a heritage museum.

Continue on Hwy 1 approximately 5 miles S to Hart's Island Rd. Turn right, cross the railroad tracks, and turn right on Robeson Rd. at the Federal Pecan Experimental Farm. Travel SW 1.9 miles on Robeson to (R) *Susanna Plantation Home*. An imposing two-story Greek Revival, good condition, beautiful grounds; private.

Return to Hart's Island Rd., turn right, and travel 1.7 miles to (L) *Gayles Plantation Home*. Greek Revival, good condition; private.

Begin in Bossier Parish at the junction of Hwy 157 and Hwy 612 (4 miles S of Haughton at Oakland) and travel W 0.3 mile on Hwy 612 to (L) *Oakland Plantation Home** (1832). Greek Revival, good condition; private.

Continue W on Hwy 612 5.4 miles to *Sligo Plantation Home* (1840). Large cottage, fair condition. Old mill, slave cabins, and other outbuildings still exist. Private.

In Webster Parish, about 6 miles E of Minden, is the junction of I-20 and LA 532. From this point travel 0.1 mile N on Hwy 532 to Webster Parish Rd. 135, and turn right to (L) *Alexander Plantation Home* (1785). Now a restaurant and antique store complex; open to public.

MONROE

At NW corner of Hwy 80 and Ingleside Dr. in NE Monroe is *Ingleside Plantation Home**. Three-story Gothic, good condition; private.

NOTE: A nominal fee is required for most homes open to the public.

* *Home included in* HISTORIC BRIEFS.

From the intersection of US 165 bypass and US 165 business, travel 4.4 miles S to Logtown School Rd. at Fondale community. Turn W and travel 0.6 mile to Logtown School. On Ouachita levee, turn left and travel 0.7 mile to (L) *Filhoil Logtown Plantation Home** (1855). Creole raised cottage, beautiful grounds; private.

Continue S on Hwy 165 from Fondale Rd. 4.6 miles to Parish Rd. 4. Turn right on gravel road and travel 0.1 mile to (R) *Bosco-Bayle Plantation Home*. Louisiana classic, poor condition; private.

On US 165, continue 6.6 miles S to entrance road (L) of *Synope Plantation Home* (1830). Creole cottage with large gallery, good condition; open to public.

Continue 2.4 miles S on Hwy 165, turn right on gravel road, and travel 0.7 mile past the grain elevator to *Breston Plantation Home* (1790). Creole cottage, good condition; private.

From the junction of Hwy 553 (Horseshoe Lake Rd.) and Parish Rd. 59, N of Monroe, travel 0.5 mile W to (L) *Glendora Plantation Home*. Raised Creole cottage, good condition, nice grounds; private.

In Marion, on Hopkins Dr. just off of Hwy 827, near the junction of Hwy 143 and Hwy 33 is *Hopkins Plantation Home** (1845). Louisiana classic, good condition; private.

In Farmerville, at 801 N. Main, is *Read Plantation Home*. Creole cottage, good condition; private.

From the junction of Hwy 33 and Hwy 2 in Farmerville, travel W 1.2 miles on Hwy 2 to (R) *Edgewood Plantation Home** (1900). Unusually beautiful Victorian design, nice grounds; private.

TALLULAH

Begin at US 80 and US 65 in Tallulah. Travel on US 80 1.9 miles to intersection with Hwy 602 (R), then go 2.7 miles right across Brush Bayou to *Crescent Plantation Home** (1832). Louisiana classic, good condition, nice grounds; private.

Continue 8.8 miles E on Hwy 602 to (R), or 1.6 miles on 602 S of junction with I-20 to *Locust Grove Plantation Home* (1890). One story with large gable roof, a gallery across the front and one side with eleven round

columns, beautiful grounds with pretty cypress bayou, and a large plantation bell on tower in yard. Good condition; private.

From the junction of Hwy 80 and Madison Parish Rd. 7 at Thomastown, 8 miles E of Tallulah, travel N 6.8 miles on Parish 7 to (L) *Dalkeith Plantation Home* (1800). Dog trot house, deteriorating badly. In 1863 this house was used by Union troops as a prison for Confederate soldiers. Private.

LAKE PROVIDENCE

On Hwy 596, travel N 0.8 mile from US 65, turn left on Schneider Ln., and travel 0.2 mile to (L) *Arlington Plantation Home** (1841). Louisiana classic, beautiful home and grounds; private.

At 600 First St. in Lake Providence is: *Amacker Plantation Home*. Creole cottage, good condition; private

From the intersection of US 65 and LA 2 NW of Lake Providence, travel 1.0 mile to (R) *Buck Meadows Plantation Home*. Modified raised cottage, fair condition; private.

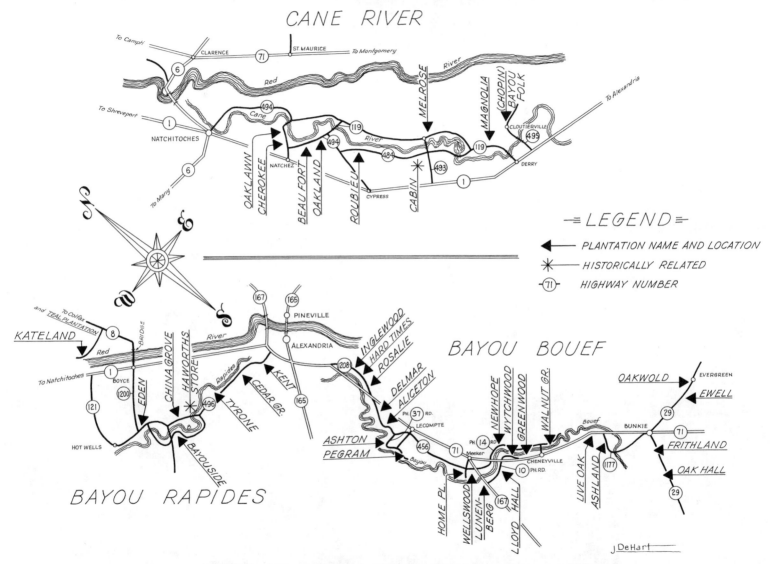

RED RIVER PLANTATION HOMES

RED RIVER PLANTATION HOMES

CANE RIVER

From the intersection of LA 1 and Williams Ave. in Natchitoches, travel 7.4 miles on Hwy 494 (Keyser Ave. in town) to *Oaklawn Plantation Home* (1840). Creole raised cottage, good condition, many trees; open to public.

Continue 0.4 mile S on Hwy 494 to *Cherokee Plantation Home* (1815). Louisiana colonial, good condition; private.

Go on 2.4 miles S on Hwy 494 and 119 to *Beau Fort Plantation Home** (1790). Creole raised cottage, good condition; open to public.

Continue 0.7 mile S on Hwy 494 and 119 to *Oakland Plantation Home** (1818). Louisiana colonial raised cottage, good condition; open by appointment.

Travel S 2.6 miles on Hwy 494 and Hwy 484 to (R) *Roubieu/Reform Plantation Home** (1808). Louisiana colonial, good condition; private.

Continue 4.6 miles S on Hwy 484. Before reaching Melrose Bridge, note the old cabin of early Louisiana bousillage and cypress stave construction. This cabin is over 200 years old. Cross the bridge to Hwy 119 and turn left to *Melrose Plantation Home** (1796). Louisiana colonial, very good condition, nice atmosphere, lovely gardens, and several outbuildings; open to public.

From Melrose Plantation, continue 4.8 miles S on Hwy 119 to *Magnolia Plantation Home** (1868). Louisiana colonial, good condition. The internationally famous thoroughbred race horse Flying Dutchman was kept at Magnolia in the 1800s and is buried on the grounds. Open by appointment.

Continue 1.3 miles to Hwy 1, go 1.5 miles S on Hwy 1 to Hwy 495, turn left to Cloutierville, and travel 0.5 mile to *Bayou Folk Museum** (1800). Louisiana colonial; open to public. This building was the home of the famous woman novelist Kate Chopin.

BAYOU RAPIDES

At the intersection of McArthur Dr. (Hwy 167) and Bayou Rapides Rd. (Hwy 496), go 0.2 mile to *Kent Plantation Home** (1800).

West Indies colonial, very good condition, several outbuildings; open to public.

Continue 4.6 miles W on Hwy 496 to *Cedar Grove Plantation Home** (1750). Louisiana colonial, very good condition, nice grounds; private.

Go 0.5 mile W on Hwy 496 to *Tyrone Plantation Home* (1830). Louisiana colonial, good condition; private.

Just W of Bayou Rapides on Hwy 496 is the Haworth Store, which has operated as a plantation country store since the mid 1800s, and which was still in service in 1979.

From Tyrone travel 4.2 miles W on 496 to *China Grove Plantation Home*. Creole cottage with three dormers, fair condition; private.

Continue 0.6 mile on Hwy 496 to *Bayou Side Plantation Home*. Creole cottage, one and one-half story. Across Bayou Rapides, this house can be seen from Hwy 496. Good condition; private.

Travel 2.7 miles on Hwy 496 to *Eden Plantation Home* (1840). Creole cottage, good condition; private.

From the junction of Hwy 8 and Hwy 158 in Colfax, travel 1.0 mile N on Hwy 158 to

(R) *Teal Plantation Home*. Modified Greek Revival, good condition; private.

From the center of the Red River bridge in Boyce, travel 4.9 miles N on Hwy 8 (east side of Red River) to an intersecting hard surface road at Kateland, turn W, and travel 1.1 miles to (R) *Kateland Plantation Home* (1870). Rambling cottage, good condition; private.

BAYOU BOEUF

Begin S of Alexandria on US 71, at the northernmost intersection of US 71 and LA 1208.1. Travel 0.9 mile on LA 1208.1 S to *Inglewood Plantation Home*. One-and-one-half-story Louisiana cottage, good condition, beautiful grounds; private.

Continue 0.5 mile on LA 1208.1 to *Hard Times Plantation Home*. Raised cottage with matching wings. Very good condition; private.

Continue 1.2 miles on LA 1208.1 S to *Rosalie Plantation Home* (1832). Beautiful one-and-one-half-story Creole cottage, beautiful grounds; private. The old sugar mill behind the home is on the National Register of Historic Places.

Travel 0.5 mile to US 71. From the intersection of US 71 and LA 470, go 0.8 mile W to *Delmar Plantation Home*. Single-story modified raised cottage, good condition; private.

Continue 1.0 mile W to *Aliceton Plantation Home*. Single-story modified raised Creole cottage, good condition; private.

From the junction of Rapides Parish Rd. 37 and Hwy 470, travel W 0.8 mile on 37 to the private road (R), and then 0.7 mile on that road to *Ashton Plantation Home* (1865). Single-story raised cottage with square pillars across front, smokehouse, wood house, and dairy, all of aged brick. Private—do not enter drive.

Continue on Hwy 37 3.6 miles to *Pegram Plantation Home*. Single story with hip roof and large gallery with square pillars across the front and half sides. Good condition; private.

From Meeker Sugar Mill travel 0.6 mile W from Hwy 71 to *Home Place/Prospect Hill Plantation Home* (1880). Greek Revival, good condition; private.

Return to LA Hwy 456 and proceed 0.5 mile S to *Wellswood Plantation Home* (1823). Louisiana classic, poor condition; private.

Travel 0.3 mile SE to Hwy 167 and then 0.9 mile S on 167 to *Lunenberg Plantation Home* (1830). Raised Creole cottage about 500 yards off to the W. Poor condition; private.

Proceed 1.1 miles E on Rapides Parish Rd. 10 to *Lloyd Hall Plantation Home** (1816). Louisiana classic, very good condition, nice

grounds, antiques, museum. Open by appointment.

From the N intersection of Rapides Parish Rd. 14 and Hwy 71, travel E 3.0 miles to *New Hope Plantation Home* (1816). Modified Louisiana classic that has been remodeled to the degree that all charm has been lost; private.

From the intersection of US 71 and Rapides Parish Rd. 39, go 0.1 mile NE on Hwy 39 to *Wytchwood Plantation Home* (1824). Greek Revival, good condition; private.

From the S junction of Parish Rd. 39 and Hwy 71, travel S 1.3 miles to *Greenwood Plantation Home* (1822). Raised cottage, good condition; private.

Go 1.0 mile S to the intersection of Hwy 71 and Main in Cheneyville. Turn E through the community and cross Bayou Boeuf to the intersection with Rapides Parish Rd. 14. Turn right and travel 1.6 miles S to *Walnut Grove Plantation Home* (1833). Louisiana classic, good condition; open to public.

Return to Hwy 71 in Cheneyville and travel 4.8 miles S to *Live Oak Plantation Home* (1854). Greek Revival, beautiful; private.

Continue 0.4 mile S on US 71, then turn onto LA Hwy 1177 and go 0.6 mile S to *Ashland Plantation Home* (1857). Greek Revival, beautiful condition; private.

Return to US 71 and the town of Bunkie.

From the intersection of US 71 and LA 29, go 5.1 miles E on LA 29 to *Ewell/Claredon Plantation Home* (1850). Modified Louisiana colonial, fair condition; private.

Travel 1.0 mile E on LA 29 to *Oakwold/Wright Plantation Home** (1835). One-and-one-half-story raised cottage, beautiful grounds, good condition; private.

In Mansura, Louisiana, on the W side of Hwy 107 (Leglise St.) in the center of town is the *Defosse Home* (1785). One-and-one-half-story Creole cottage with wide porch and eight slim cypress columns. Good condition considering age; private.

Located on the N side of Hwy 1, next to the post office in Hamburg, Louisiana, which is E of Mansura, is *Callihan Plantation Home** (1841). Large raised Creole cottage with square columns; private.

In Moreauville, located on the N side of Main St. three houses E of Bayou des Glaises St., is *Lougaure Plantation Home*. One-and-one-half-story Creole cottage, good condition; private.

Return to Bunkie and proceed 1.0 mile S from the intersection of US 71 and LA 29 to *Frithland Plantation Home**. Louisiana classic, very good condition, nice grounds; private.

Continue 3.8 miles on Hwy 29 to *Oak Hall Plantation Home*. Louisiana classic, beautiful. Cannot be seen from road, but no trespassing is allowed; private.

From the junction of Hwy 29 and Hwy 114 SE of Moreauville, travel 1.7 miles S on Hwy 29 to (L) *Bordelon Plantation Home* (1870). Louisiana Creole cottage with spacious galleries on three sides, good condition; private.

On Hwy 29 near Bordelon plantation home is *Cappel Plantation Home* (1890). Louisiana Creole cottage with large dormer extended from one side only. Under repairs; private.

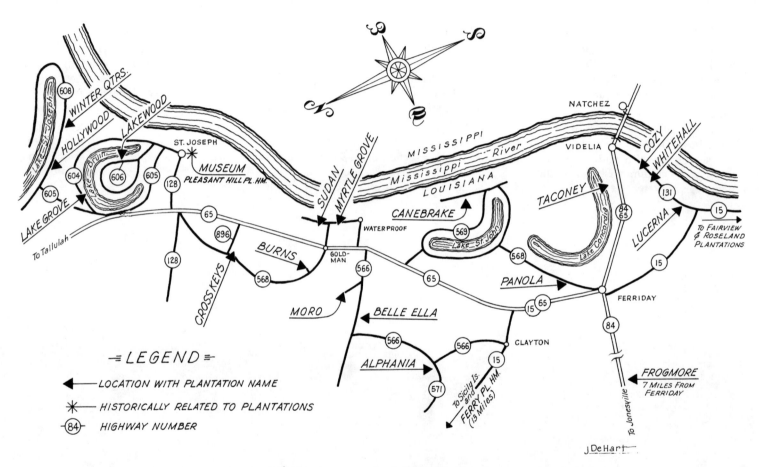

· TENSAS PLANTATION HOMES ·

TENSAS PLANTATION HOMES

From the junction of Hwy 84 and Hwy 131 in Vidalia (at the foot of the Mississippi River bridge), travel 2.3 miles S on Hwy 131 to (R) *Cozy Plantation Home*. Raised Creole cottage, good condition, nice grounds; private.

Continue S 0.7 mile on Hwy 131 to (R) *Whitehall Plantation Home* (1800). Creole cottage (modernized), good condition, nice grounds, large plantation bell on front lawn, private.

Continue S 3.0 miles to (R) *Lucerna Plantation Home*. Creole raised cottage in fair condition, magnificent oak trees near house; private.

On Hwy 84, go 2.0 miles W of the intersection with Hwy 131 in Vidalia (about 300 yards N of Hwy 84 on side road) to *Taconey Plantation Home* (1802). Modified Louisiana colonial, being renovated. Will be open to public when complete.

From the junction of Hwy 84 and Hwy 568 in Ferriday, travel 1.6 miles on Hwy 568 to (L) *Panola Plantation Home*. Creole cottage, good condition; private.

Continue on Hwy 568, turn right at the junction with Hwy 569, and travel E. From this point travel 0.7 mile on Hwy 569 to the hard surface road, turn right, and travel 0.9 mile N to (L) *Canebrake Plantation Home*. Raised Creole cottage, good condition. The complex surrounding includes an old church, barns, and cabins. Private.

Return to Ferriday. On Hwy 84, 7.0 miles W of the junction with Hwy 15 in Ferriday, is *Frogmore Plantation Home* (1820). Creole raised cottage, very good condition; private.

From the junction of Hwy 8 and Hwy 1017 in Sicily Island (Catahoula Parish), travel 1.3 miles S on Hwy 1017 to (R) *Ferry Plantation Home** (1830). Louisiana classic, good condition, nice grounds; private. General store, barns, dovecotes, and Indian mounds are in the yard, overlooking Lake Louis.

Go to the intersection of Hwy 15 and Hwy 566 in Clayton and travel 3.4 miles N on Hwy 566 to the intersection with Hwy 571 to (L) *Alphania Plantation Home*. Louisiana colonial, badly deteriorated; private.

Continue on Hwy 566 5.7 miles to the intersection with Hwy 571 N. Turn E on Hwy 566 and travel 1.2 miles to (R) *Belle Ella Plantation Home*. Creole raised cottage, fair condition; private.

Continue E on Hwy 566 1.6 miles to the entrance road (L) of *Moro Plantation Home*. Louisiana colonial, fair condition; private.

On Hwy 568 at Goldman (N of Waterproof), travel E 1.2 miles, and turn left 0.1 mile along levee road (L) to *Sudan Plantation Home*. Louisiana classic, good condition; private.

Return to Hwy 568 and travel S 0.1 mile along levee road (R) to *Myrtle Grove Plantation Home* (1810). This Louisiana raised cottage was prefabricated in Louisville, Kentucky, shipped downriver on barges, and constructed. Beautifully refurbished; open to public.

From the junction of Hwy 568 and Hwy 65 N of Waterproof, travel NW 1.1 miles on Hwy 568 to entrance road (R) of *Burns Plantation Home** (1853). One-and-one-half-story Creole cottage in good condition, still operating as a plantation; private.

Continue N 4.7 miles on Hwy 568 to the junction with Hwy 896 (R) to *Cross Keys Plantation Home*. Two-story Creole cottage, good condition. The house can be better seen from Hwy 896. Slave cabins are still in use and the plantation is still operating; private.

NOTE: A nominal fee is required for most homes open to the public.

Return to Hwy 65 on Hwy 896 and turn right to the town of St. Joseph. In town, there are several interesting buildings, including *Christ Episcopal Church*, built in 1872; open to public. *Christ Church Rectory*, also known as Bondurant House, was built before 1852; open to public.

A plantation museum is housed on Plank St. in *Pleasant Hill Plantation Home*, which was moved to this location. Louisiana colonial, very good condition; open to public.

From the junction of Hwy 604 and Hwy 606 at Lake Bruin (N of St. Joseph), travel 0.7 mile on Hwy 606 to a fork in the highway. Take the road on right and travel 0.9 mile along the inner circle of Lake Bruin to (R) *Lakewood Plantation Home** (1854). Large one-and-one-half-story raised Creole cottage facing Lake Bruin, very good condition; private.

At the community of Lake Bruin on US 65 turn E onto Hwy 607. At the intersection with Hwy 605 travel N 0.6 mile to the junction with Hwy 604, turn right on Hwy 604, and travel 1.7 miles E to (R) *Lake Grove Plantation Home*. Louisiana colonial, good condition, beautiful grounds; private.

Return to Hwy 605, turn right (N), and travel 2.8 miles to intersection with Hwy 608. Turn right and travel 0.6 mile to (R) *Hollywood Plantation Home*. Creole cottage, good condition; private.

Continue on Hwy 608 2.8 miles to (R) *Winter Quarters Plantation Home** (1852). Modified Louisiana colonial, fair condition; open to public.

Two additional plantations are located on a very poor road. Because of the condition of the homes and the road the trip probably is not worthwhile. For those that are interested, start from the junction of Hwy 15 and Hwy 565 and turn E onto levee road. Travel S 9.1 miles to side road and then 1.6 miles to (R) *Fairview Plantation Home*. Cottage, in state of deterioration; private.

Continue 3.0 miles to *Roseland/Ashland Plantation Home*. Creole cottage, fair condition; private.

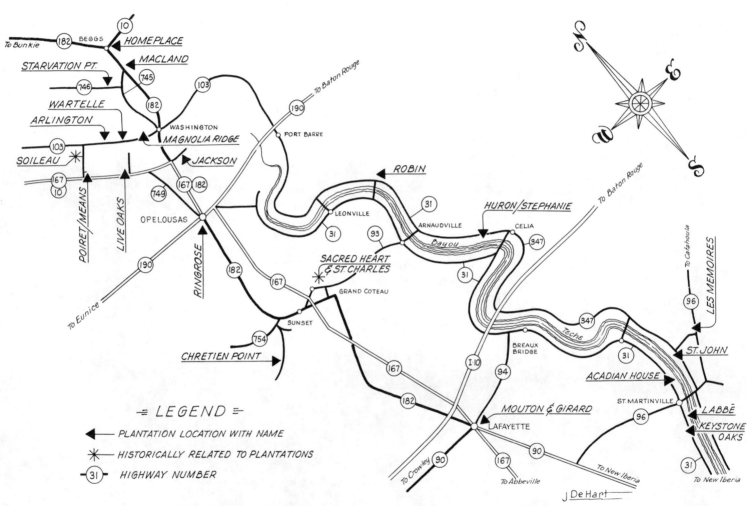

• ATTAKAPAS PLANTATION HOMES •

LAFAYETTE

At 338 N Sterling Rd. in Lafayette is *Mouton Plantation Home** (1848). Louisiana colonial, good condition; private.

Located at 1340 Pinhook Rd., now serving as the Old Acadian Inn, in Lafayette, is *Girard Plantation Home** (1820). Louisiana classic, good condition; open to the public.

From the junction of LA 347 and Hwy 96 in St. Martinville, travel N on LA 347 2.5 miles to: *St. John Plantation Home** (1828). Greek Revival, excellent condition, beautiful grounds and oaks; private.

On St. John Rd., off of Hwy 347 directly opposite St. John plantation, travel 1.5 miles to Hwy 96, turn left (E), and travel 0.7 mile to (L) *Les Memoirs Plantation Home** (1836). Louisiana colonial, good condition; private.

Return to St. Martinville on Hwy 96 and turn left to Hwy 31 and Gary St. From this point travel S 1.2 miles on Hwy 31 to (L) *Labbe Plantation Home** (1850). Large Louisiana Creole cottage, good condition; private.

Continue 1.7 miles S on Hwy 31 to (R) *Keystone Oaks Plantation Home** (1870). Louisiana classic, good condition; private.

Return to Hwy 96 and Hwy 31 in St. Martinville and travel N 1.0 mile on Hwy 31 to Longfellow Park and *Acadian House Museum/ Arceneaux Plantation Home** (1765). Modified Louisiana colonial, good condition; open to public.

From Bayou Teche bridge and Hwy 347 in Cecilia, travel 2.0 miles N on 347 to *Huron/ Stephanie Plantation Home* (1850). Louisiana colonial, sunbaked brick house, beautiful architecture, badly in need of repair; private.

Travel Hwy 347 to Hwy 31 14.0 miles to *Robin Plantation Home* (1830). One-and-one-half-story cottage, built by the son of one of Napoleon's generals. Fair condition; private.

Located in Grand Coteau is *Sacred Heart Academy**. Established in 1821, this was a school for planters' daughters in the Acadian Country. The nearby *St. Charles Academy*, was established in 1838 for the education of planters' sons. It is now a Jesuit seminary.

From the junction of Hwy 754 and Hwy 182, 2.0 miles N of Sunset, travel on Hwy 754 0.2 mile W to side road, turn left, and then, bearing constantly to the right, continue 2.0 miles to *Chretien Point Plantation Home** (1830). Louisiana classic, good condition; open to public.

In Opelousas, at 1152 Prudhomme Circle, off Prudhomme Ln., is *Ringrose Plantation Home** (1770). Louisiana colonial, soon to be restored and opened as a museum.

From the intersection of Hwy 749 and Hwy 167, N of Opelousas, travel 0.2 mile W on Hwy 167 to a side gravel road (R), then 0.6 mile N. On the left, about 600 yards from the road in a grove of oaks, is *Live Oaks Plantation Home*. Louisiana colonial, fair condition; private.

Return to Hwy 167, turn right (W), and continue 3.5 miles to a hard surface road. Turn right and travel 0.3 mile to entrance (R) of *Poiret/Means Plantation Home**. Louisiana colonial, very good condition; private.

From the junction of LA 182 and LA 103 at Washington, go 0.3 mile W on Hwy 103 to *Magnolia Ridge Plantation Home** (1830). Louisiana classic, good condition; open by appointment.

Continue 1.5 miles on Hwy 103 to *Wartelle Plantation Home* (1827). Raised Creole cottage, good condition; private.

Continue 0.6 mile W on Hwy 103 and then 1.0 mile N on the private road to *Arlington Plantation Home* (1829). Modified Greek Revival, good condition; open to public.

Return to Hwy 103 and proceed 2.6 miles W on Hwy 103 to *Soileau Home*. Although this is not a plantation home, it is an example of unusual French Creole design, and is one of the oldest homes in Louisiana. Well preserved. Notice the many Acadian homes in this area.

Return to Hwy 182 at Washington. Continue 3.0 miles N on 182 to *Macland Plantation Home** (1840s). Louisiana classic, poor condition, badly neglected; private.

At Macland, from Hwy 182, travel 0.8 mile W on Hwy 745 to Hwy 746 and then 0.2 mile N on Hwy 746 to *Starvation Point Plantation Home** (1790). Red brick Louisiana classic, good condition, barren grounds; private.

Return to Hwy 182 and travel N to Beggs; then turn left on Hwy 182 to (R) *Homeplace Plantation Home*. Louisiana raised cottage, good condition, nice grounds; open to public.

In addition, there are in Opelousas and in Washington several homes and structures dating to the great plantation era. Some of these buildings are of substantial historical interest.

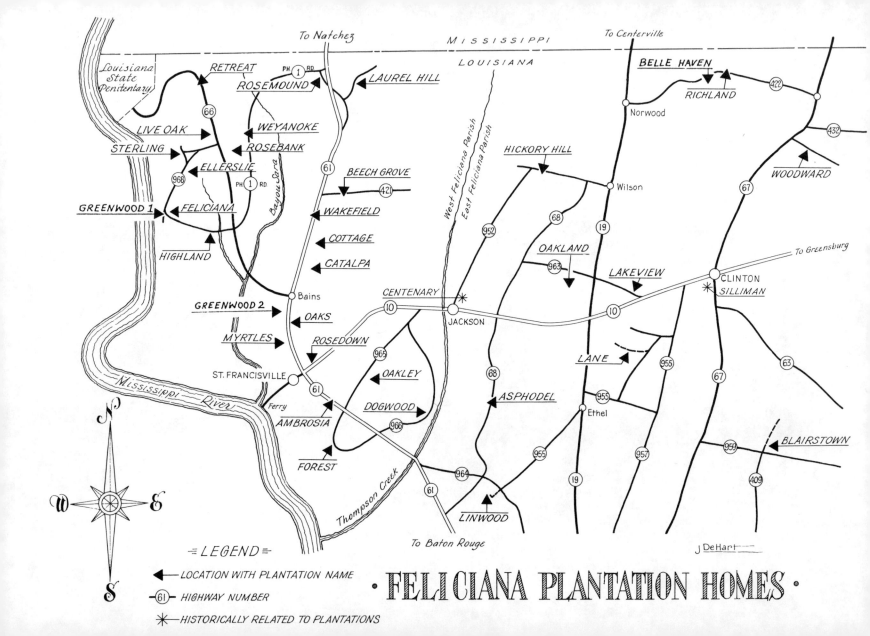

FELICIANAS PLANTATION HOMES

ST. FRANCISVILLE

From the junction of Hwy 964 and Hwy 955 NW of Zachary, travel W 0.8 mile on gravel road to the entrance of *Linwood Plantation Home** (1838). Louisiana classic, good condition; private.

On Hwy 68, from the intersection with Hwy 61, travel NE 7.2 miles to entrance road (L) of *Asphodel Plantation Home** (1835). Louisiana classic, good condition; open to public.

At the intersection of Hwy 61 and Hwy 966 SE of St. Francisville, turn S on Parish Rd. 7 and travel 2.1 miles to (L) *Forest Plantation Home** (1894). Rambling cottage, good condition; private.

At the intersection of Hwy 61 and Hwy 966 SE of St. Francisville, turn N on Hwy 966 and travel 3.4 miles N to (L) *Dogwood Plantation Home** (1803). Louisiana Creole cottage, newly renovated; private.

On Hwy 965, from the intersection with Hwy 61, travel NE 2.8 miles to road (R) to *Oakley Plantation Home** (1810). Louisiana colonial, good condition. Site of Audubon Memorial Park. Open to public.

From the junction of Hwy 61 and Hwy 965, travel 0.2 mile W on Hwy 61 to (R) *Ambrosia Plantation Home*. Raised Creole cottage, good condition, nice grounds; private.

On Hwy 10, from the intersection with Hwy 61, travel E 0.3 mile to entrance road (L) of *Rosedown Plantation Home** (1835). Louisiana classic, good condition, beautiful grounds; open to public.

On Hwy 61, from the intersection with Hwy 10, go N 1.2 miles to entrance road (L) of *Myrtles Plantation Home** (1830). Creole raised cottage, good condition, nice grounds; open to public.

Continue 0.2 mile on Hwy 61 to entrance road (R) of *The Oaks Plantation Home*. Creole raised cottage, good condition; private.

Go 0.5 mile farther on Hwy 61 to entrance road (L) of *Greenwood II Plantation Home* (1779). Louisiana classic, good condition; private. This home cannot be seen from the highway.

Continue N on Hwy 61 for 1.0 mile to Bains. From the intersection with Hwy 66, stay on Hwy 61 for 1.2 miles to entrance road (R) of *Catalpa Plantation Home* (1835). Creole raised cottage, good condition; open to public.

Return to Hwy 61 and proceed 1.0 mile

to entrance road (R) of *Cottage Plantation Home** (1811). Creole cottage, good condition; open by appointment.

Return to Hwy 61 and continue 2.5 miles N to entrance road (L) of *Wakefield Plantation Home* (1833). Creole raised cottage, good condition; private.

From Hwy 61, turn E onto Hwy 421 and go 1.2 miles to (R) *Beech Grove Plantation Home*. Louisiana Creole cottage, good condition; private.

From the Mississippi State Line, travel 1.0 mile S on Hwy 61 to (R) Sligo Rd. (Parish Rd. 1). Turn W, travel 0.6 mile, and turn left on old Parish Rd. 3 (Old Hwy 61) and travel 0.1 mile to (R) *Rosemound Plantation Home*. Victorian, good condition.

Continue on Parish Rd. 3 to the junction with Hwy 61. Cross over to old Laurel Hill Road and travel 0.6 mile to Harris Corner Rd. Turn left and go 1.2 miles to entrance road (R) of *Laurel Hill Plantation Home** (1830). Modified Louisiana classic, good condition; private.

Return to Bains and the intersection of Hwy 66 and Hwy 61. Go NW 7.5 miles on Hwy 66 to entrance road (R) of *Rosebank*

NOTE: A nominal fee is required for most homes open to the public.

* Home included in HISTORIC BRIEFS.

*Plantation Home** (1790s). Louisiana colonial, fair condition; private.

Continue NW 0.8 mile to entrance road (L) of *Live Oak Plantation Home* (1802). Louisiana colonial, good condition; private.

Continue NW 3.1 miles to (L) *Retreat Plantation Home* (1850). Modified Louisiana colonial raised cottage, good condition; private.

Return SE 3.4 miles on Hwy 66 to the intersection with Hwy 968. Turn right and continue 0.7 mile to the fork in the road (R) to *Sterling Plantation Home*. Modified Louisiana colonial, good condition; private.

Return to the fork in road and continue W 0.3 mile (L) to *Ellerslie Plantation Home** (1832). Louisiana classic, good condition; private.

Continue W on asphalt road for 1.8 miles. Practically hidden by trees to the left is the rear of *Feliciana Plantation Home* (1830). Louisiana classic, good condition; private. Continue 0.1 mile (R) to *Greenwood I Plantation Home* (1830). This home has been rebuilt to exact specifications of the original home-- beautifully done. Greek Revival; open to public.

Continue 2.9 miles E on asphalt road to *Highland Plantation Home** (1805). Louisiana

classic, good condition, beautiful grounds; private.

Continue E 0.6 mile to the intersection with Hwy 66. On the E side of Hwy 66 enter West Feliciana Parish Rd. 1 to Hollywood. Travel 4.3 miles on Parish Rd. 1 to (L) *Weynoke/Beechwood Plantation Home** (1809). Modified Louisiana classic, good condition; private.

CLINTON

Travel on Hwy 963 1.8 miles E of the intersection with Hwy 19 to *Lakeview Plantation Home* (1840). Two-story Louisiana classic with front gallery and an old kitchen that is still in use. Very good condition, nice grounds; private.

Continue on Hwy 963 1.4 miles W of the intersection with Hwy 19, near Gurley, to *Oakland Plantation Home* (1827). Louisiana classic, good condition, beautiful double row of oaks at the entrance; private.

From the intersection of Hwy 963, travel 1.2 miles N on Hwy 68 to *Glencoe Plantation Home* (1870). Victorian Gothic, very good condition and beautifully renovated; open to public.

Near the town of Wilson, 1.7 miles W from intersection of Hwy 68 and Hwy 952 on Hwy 952 is *Hickory Hill Plantation Home* (1810). Greek Revival, fair condition, grounds in fair condition; private.

On Hwy 422 4.9 miles E of the intersection with Hwy 19, at Norwood, is *Richland Plantation Home* (1820). Greek Revival, very good condition, beautiful grounds; private.

Continue W 3.6 miles to *Belle Haven Plantation Home* (1830). Creole cottage, very good condition; private.

From the junction of Hwy 432 and Hwy 67, approximately 8 miles NE of Clinton, travel 1.3 miles SW on Hwy 67 to road on left. Turn SE onto this road and travel 0.8 mile to *Woodward/Hollygrove Plantation Home* (1852). Modified Greek Revival, two story brick, fair condition; private.

Go 4.8 miles E on Hwy 959 from the intersection with Hwy 67. At the intersection with Hwy 409, proceed 0.1 mile N on gravel road to *Blairstown Plantation Home* (1850). Greek Revival, fair condition, grounds fair condition; private.

From the junction of Hwy 10 and Hwy 955 W of Clinton, travel S on Hwy 955 3.1 miles to asphalt road. Turn right, travel 0.5 mile, and turn left on asphalt road. Drive 0.8 mile S to an asphalt road, turn right, and go 0.5 mile to (L) *Lane Plantation Home** (1825). Modified Louisiana classic, good condition; private.

Returning to the asphalt road, you may wish to turn right for the beautiful scenic forest drive to Hwy 955. Or drive on Hwy 952 just N of Jackson to *Centenary College*, which was established in 1825 to educate the plantation youth of the Felicianas. The Methodist Conference moved the college to Shreveport in 1908; only part of the building now remains.

In SW Clinton *Silliman College* may be seen. Established in 1852 to educate the youth of the plantation area, Silliman operated until 1931.

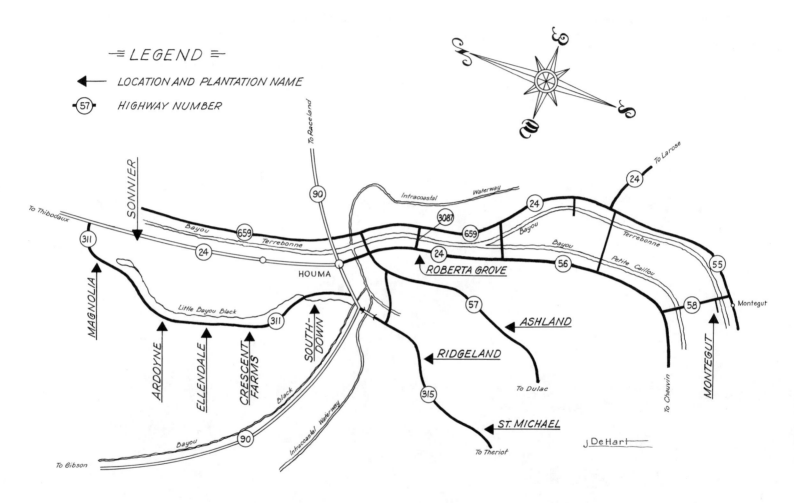

· TERREBONNE PLANTATION HOMES ·

LEGEND

← LOCATION AND PLANTATION NAME

(57) HIGHWAY NUMBER

J. DeHart

TERREBONNE PLANTATION HOMES

From the junction of Hwy 24 and Hwy 311 N of Houma, travel 1.0 mile S on Hwy 24 to (R) *Sonnier Plantation Home* (1800). Louisiana classic, fair condition; private.

Return to the junction of Hwy 311 and Hwy 24 and take Hwy 311 S 1.3 miles to (R) *Magnolia Plantation Home* (1858). Louisiana colonial, fair condition; private.

Continue S 3.4 miles to (R) *Ardoyne Plantation Home** (1890s). Very ornate Gothic, good condition; open by appointment.

Go W 2.3 miles to *Ellendale Plantation Home* (1810). Victorian cottage, good condition; private.

Continue S 1.7 miles to (R) *Crescent Farms**

(1834). Louisiana colonial, good condition; private.

Proceed S 3.5 miles to (R) *Southdown Plantation Home** (1858). Large Gothic, fair condition; open to public.

From the junction of Hwy 24 and Hwy 57 on Bayou Terrebonne in Houma, travel 5.3 miles S on Hwy 57 to (L) *Ashland Plantation Home.* Large one-and-one-half-story Creole cottage, good condition, nice grounds; private.

On the SW corner of Prospect Ave. and E Main St. (the junction of Hwy 24 and Hwy 3087 at Bayou Terrebonne bridge) is *Roberta Grove Plantation Home.* Victorian design, beautiful grounds; private.

Beginning at the center of the Intracoastal Waterway bridge on Hwy 315 in SW Houma, travel 4.1 miles S to (L) *Ridgeland Plantation Home* (1850). Large Creole cottage reconstructed to the exact design of the original plantation home. Good condition; private.

Continue 4.4 miles S on Hwy 315 to (L) *St. Michael Plantation Home.* Large Creole cottage, good condition; private.

On Aragon Rd., S of Hwy 58 on the W bank of Bayou Terrebonne in Montegut (S of Houma), is *Montegut Plantation Home* (1839). Modified Gothic with slave cabins and remains of an old brick sugar mill. Good condition; private.

NOTE: A nominal fee is required for most homes open to the public.

* Home included in HISTORIC BRIEFS.

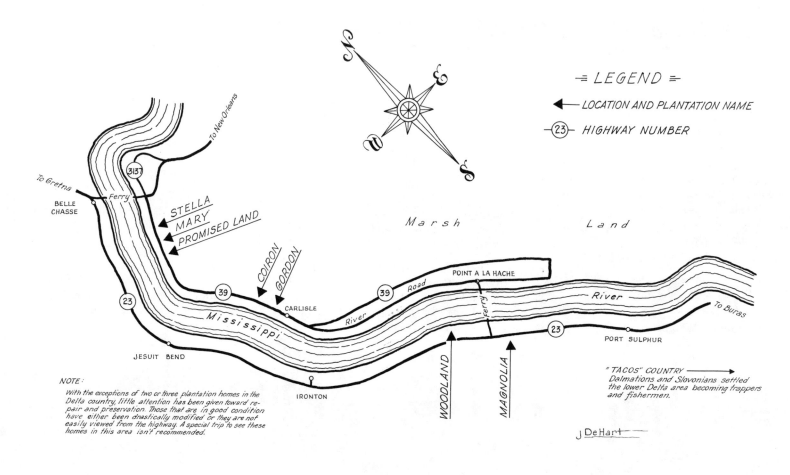

N

W E

S

≡ LEGEND ≡

← LOCATION AND PLANTATION NAME

-(23)- HIGHWAY NUMBER

To New Orleans

(3I37)

To Gretna

Ferry

BELLE CHASSE

STELLA
MARY
PROMISED LAND

COIRON
GORDON

Marsh   Land

POINT A LA HACHE

(23)

(39)

(39)   Road

River

Ferry

River

To Buras

(23)

Mississippi

CARLISLE

River

PORT SULPHUR

WOODLAND

MAGNOLIA

JESUIT BEND

IRONTON

NOTE:

With the exceptions of two or three plantation homes in the
Delta country, little attention has been given toward repair and preservation. Those that are in good condition
have either been drastically modified or they are not
easily viewed from the highway. A special trip to see these
homes in this area isn't recommended.

"TACOS" COUNTRY →
Dalmations and Slovonians settled
the lower Delta area becoming trappers
and fishermen.

J DeHart

• DELTA PLANTATION HOMES •

DELTA AREA PLANTATION HOMES

With the exception of two or three plantation homes in the Delta country, little attention has been given to repair and preservation of the historic dwelling places. Those that are in good condition either have been drastically modified or are not easily visible from the highway. For these reasons, a special trip to this area may not be worthwhile to some travelers.

EAST BANK

Continue 2.1 miles S on Hwy 39 to *Stella Plantation Home*. Modified Greek Revival with six columns, beautiful oaks and grounds. Good condition; private.

Drive 1.2 miles S on Hwy 39 to *Mary Plantation Home** (late 1700s). Louisiana colonial, very good condition; private. This home cannot be easily viewed from the road.

Continue 0.5 mile S on Hwy 39 to *Promised Land Plantation Home* (late 1700s). Louisiana Creole raised cottage, good condition; private.

Proceed 8.8 miles on Hwy 39 to *Coiron Plantation Home*. Creole raised cottage, fair condition; private.

Continue 0.3 mile S to *Gordon Plantation Home* (1838). Creole cottage that is now the Carlisle Post Office. Private.

WEST BANK

Start at West Pointe a la Hache. From the ferry road on Hwy 23 (west bank River Rd.) travel SE 1.3 miles to *Magnolia Plantation Home** (1795). On the right side of the road, almost in complete ruins, stand three of the main buildings. Across the road is the old Magnolia Plantation store. Although the store is in a deteriorated condition, it is not beyond repair; it could be an interesting and valuable landmark if properly restored. Private.

From the ferry road in West Pointe a la Hache travel NW on Hwy 23 2.0 miles to *Woodland Plantation Home** (early 1800s). Louisiana colonial located on the right side of the road about 300 feet toward the river. It is badly in need of repair, but could be beautifully restored without losing the original appearance. Private.

NOTE: A nominal fee is required for most homes open to the public.

* Home included in HISTORIC BRIEFS.

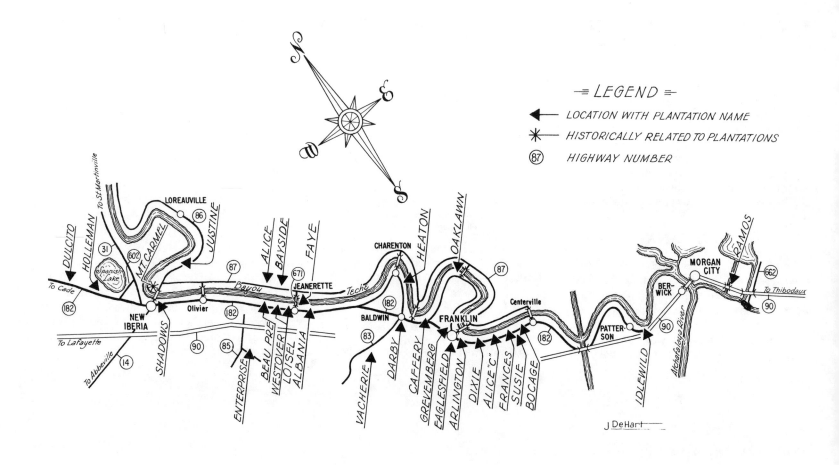

J DeHart

· TECHELAND PLANTATION HOMES ·

TECHELAND PLANTATION HOMES

In New Iberia, on Main St. (Hwy 182) at Weeks St., is the home *The Shadows on the Teche** (1831). Louisiana classic, good condition; open to the public.

From the junction of Hwy 677 and Hwy 182 in New Iberia, travel W 1.9 miles on Hwy 182 to (R) *Holleman/Segura Plantation Home** (1812). Louisiana colonial, good condition; private.

Continue W 2.3 miles on Hwy 182 to (R) *Dulcito Plantation Home* (1788). Creole cottage, good condition. House cannot be seen from highway because of the trees. Open by appointment only—check with the New Iberia Tourist Center.

Return to New Iberia. From Main St. (Hwy 182) cross Bayou Teche at Duperier Ave. to (L) *Mount Carmel Convent* (1872). This institution was established by the Sisters of Mount Carmel for the education of young girls in the Teche area.

Continue NE on Duperier Ave. 2.0 miles to (L) *Justine Plantation Home* (1822). Creole cottage, good condition; open to public by appointment.

Located at Patoutville on Hwy 673, 1.5 miles SW of the intersection with Hwy 90 near Jeanerette, is *Enterprise Plantation Home** (1850). Louisiana classic, good condition; private.

From the junction of Hwy 85 and Hwy 182, travel 0.5 mile W on Hwy 182 to (R) *Westover Plantation Home* (1860). Modified two-story Louisiana classic, good condition; private.

Continue 0.8 mile W on Hwy 182 to (R) *Loisel Plantation Home* (1830). Large Creole cottage, good condition; private.

Continue 1.3 miles W on Hwy 182 to (R) *Beau Pre Plantation Home** (1830). Louisiana colonial, good condition; private.

Return to Jeanerette. At 217 Lewis St. is *Faye Plantation Home** (1857). Creole cottage, good condition; private.

From Lewis St. in Jeanerette, cross Bayou Teche to Hwy 87, turn left, and travel 1.6 miles W to *Bayside Plantation Home** (1850). Louisiana colonial, good condition; private.

Continue 0.5 mile W on Hwy 87 to *Alice Plantation Home** (1800). Louisiana colonial, good condition. This is the Fuselier Plantation Home, formerly located near Baldwin. It was moved by barge to this location.

Return to Hwy 182 in Jeanerette and travel 1.0 mile E of Lewis St. to *Albania Plantation Home** (1842). Louisiana classic, three stories, good condition; open to public.

Continue 8.0 miles to Baldwin. Just opposite the junction of Hwy 182 and Hwy 83 (L), now housing a bank, is *Darby Plantation Home** (1765). Louisiana colonial, good condition; private.

From the junction of Hwy 182 and Hwy 83 in Baldwin, travel Hwy 83 W 4.0 miles to (L) *Vacherie Plantation Home* (1815). Louisiana colonial; private. The home cannot be seen from the highway, but it is on posted property.

From the junction of Hwy 182 and Hwy 83, travel E on Hwy 83 2.6 miles to (R) *Heaton Plantation Home* (1853). Italian villa, good condition; private.

Return to Hwy 182 at Baldwin and travel E 2.5 miles to (L) *Caffery Plantation Home*. Louisiana Creole cottage; private. The building can barely be seen through a heavy grove of trees.

Immediately after passing the Caffery Home, take the hard surface road on left and travel 3.5 miles NE to (L) *Oaklawn Plantation*

NOTE: A nominal fee is required for most homes open to the public.

Home* (1827). Greek Revival, good condition, beautiful grounds; open by appointment.

On Sterling Rd. in NE Franklin, is the St. Mary Museum, *Grevemberg House* (1851). Greek Revival; open to public.

In Franklin at Main St., near the Catholic church, turn S on Iberia and go to Anderson St. Turn left to Hansen High School; on side street go to rear of school (L) to *Eaglesfield Plantation Home* (1859). Louisiana classic, good condition; private. The building is now used as the home economics building of Hansen High School.

From the junction of Hwy 182 and Bridge St. in Franklin, travel 1.1 miles E on Hwy 182 to *Arlington Plantation Home** (1850). Greek Revival, good condition; private.

Continue 0.4 mile E on Hwy 182 to *Dixie Plantation Home** (1850). Greek Revival, good condition; private.

Proceed 0.8 mile E on Hwy 182 to *Alice C Plantation Home* (1847). Louisiana classic, good condition; private.

Continue 0.9 mile E on Hwy 182 to *Frances Plantation Home** (1820). Louisiana colonial, good condition; open to public.

Go on 0.6 mile on Hwy 182 to (R) *Susie*

Plantation Home* (1852). Louisiana classic, good condition; private.

Continue 0.9 mile E on Hwy 182 to *Bocage Plantation Home** (1846). Greek Revival, good condition; open to tours.

At the junction of Hwy 182 and 90 E Patterson is *Idlewild Plantation Home** (1850). Creole raised cottage, good condition; private.

Travel E to the Atchafalaya River bridge at Morgan City. From the center of the bridge span travel on Hwy 90 E 5.4 miles to (L) *Ramos Plantation Home* (1854). Louisiana Creole cottage, good condition; private.

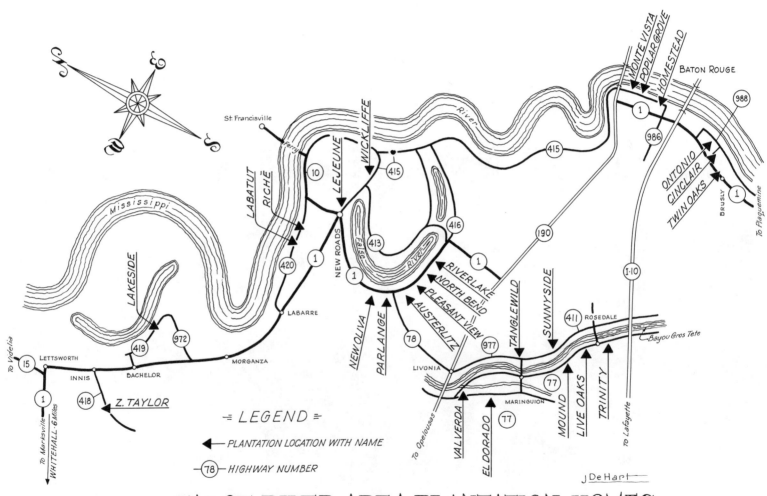

FALSE RIVER AREA PLANTATION HOMES

LEGEND

◄ PLANTATION LOCATION WITH NAME

(78) HIGHWAY NUMBER

J DeHart

FALSE RIVER AREA PLANTATION HOMES

FALSE RIVER

Begin at the intersection of Hwy 416 and LA 1 at False River. Travel 0.5 mile NW on Hwy 1 to *River Lake Plantation Home*. Louisiana colonial, including slave cabins and dovecotes. Fair condition; private.

Continue 1.0 mile on LA 1 to *North Bend Plantation Home*. Louisiana colonial, fair condition; private.

Proceed N on Hwy 1 0.5 mile to *Pleasant View Plantation Home*. The building is hard to see from the road. Good condition; private.

Continue N on Hwy 1 1.0 mile to *Austerlitz Plantation Home** (1832). Louisiana colonial, good condition; private.

Go N 1.4 miles to *Parlange Plantation Home** (1750). Louisiana colonial, good condition; open to public.

Continue 1.0 mile to *New Oliva Plantation Home* (1850). Louisiana colonial, good condition; private.

Continue 5.1 miles to the corner of Hwy 1 and Lejeune St. in the town of New Roads, to find *Lejeune Plantation Home*. Louisiana colonial, good condition; open by appointment.

Travel 2.8 miles from Lejeune on Hwy 415 to *Wickliffe Plantation Home*. Raised cottage, good condition; private.

Return to Hwy 10 on Hwy 1 and then travel 2.3 miles to the intersection with Hwy 420. Turn left, get on River Rd. (Hwy 420), and go 2.5 miles to *Riche Plantation Home** (1825). Raised Creole cottage, good condition; private.

Continue 0.5 mile W to *Labatut Plantation Home** (1790). Louisiana colonial, in a state of deterioration; private.

From the junction of Hwy 1 and Hwy 419 at Bachelor, go SE on River Rd. (Hwy 419) 1.2 miles to *Lakeside Plantation Home* (1860). Raised cottage with cast iron grillwork posts and beautiful grounds. Good condition; private.

From the junction of Hwy 1 and Hwy 418 at Innis, turn W on Hwy 418 and travel 2.3 miles to *Zachary Taylor Plantation Home**. Two-and-one-half-story Louisiana classic with six square columns from ground to roof. Good condition; private.

From Lettsworth travel W approximately 5 miles to the E bank of the Atchafalaya River at Legionaire. From the junction of Hwy 417 and Hwy 418 N of Hwy 1, proceed 1.1 miles N on River Rd. (Hwy 418) to (R) *Whitehall Plantation Home**. Louisiana classic, good condition; private.

BAYOU GROSSE TETE

Begin at the intersection of I-10 and Hwy 77. Go 1.6 miles N on Hwy 77 to *Trinity Plantation Home** (1839). Louisiana colonial led to by an avenue of live oaks. Good condition; private.

Continue N on Hwy 77 0.6 mile to *Live Oaks Plantation Home** (1828). Two-and-one-half-story Louisiana colonial house with six square columns. The large "Mays Oak," a small brick slave church, servants' buildings, and old kitchen are on grounds. Good condition; open by appointment.

Proceed 2.6 miles N on Hwy 77 to *Mound Plantation Home** (1840). One-story frame house built on an old Indian mound with a brick slave laundry to the rear. Fair condition; private.

Continue N on Hwy 77 to Maringouin, turn right, and cross the bridge to *Tanglewood Plantation Home* (1850). One-story cottage set in a grove of live oaks; private.

Travel 2.0 miles SE on Hwy 411 to (L)

NOTE: A nominal fee is required for most homes open to the public.

* Home included in HISTORIC BRIEFS.

Sunnyside Plantation Home (1836). Louisiana classic, beautiful grounds. Good condition; private.

Return to the bridge at Tanglewood. From this point travel 3.5 miles NW on Hwy 77 to *Eldorado Plantation Home* (1850). Raised cottage with two-story brick slave quarters to the rear. Good condition; private.

Continue N on Hwy 77 1.3 miles to LA 977. Turn right and go 0.5 mile to *Valverda Plantation Home*. Two-story brick Greek Revival with six white columns. Good condition; private.

BRUSLEY

Just NE of the intersection of LA 1 and Hwy 987.2 at Brusley, near Plaquemine, is *Twin Oaks Plantation Home*, which was moved to this location from Plaquemine. Louisiana colonial, very good condition, grounds and gardens in excellent shape; private.

Continue N 0.6 mile on Hwy 1 and turn right on Cinclaire Plantation Rd. to *Cinclaire Plantation Home*. The grounds feature an old mill. Good condition; private.

Continue to Hwy 988 (River Rd.), turn left, and travel 0.6 mile to *Antonio Plantation Home* (early 1800s). Louisiana colonial, good condition; private.

PORT ALLEN

From the intersection of Hwy 987.1 and River Rd. (Hwy 986) beneath the W approach of the Hwy 190 Mississippi River bridge at Port Allen, travel S 0.4 mile on Hwy 986 to *Monte Vista Plantation Home* (1850). Louisiana colonial, good condition; private.

Continue S 0.6 mile on River Rd. to *Poplar Grove Plantation Home* (1884). This home of very unusual style was part of the Bankers' Pavilion of the New Orleans Cotton Centennial in 1884 and was moved upriver by Horace Wilinson, Sr., as the center of a plantation. The slave cabins for Poplar Grove, 0.3 mile S of the house, are still in good condition. The old sugar mill is still standing. Private.

From the intersection of Hwy 1 and Hwy 986 in Port Allen, travel NE 0.6 mile to *Homestead Plantation Home*. Greek Revival, excellent condition; private.

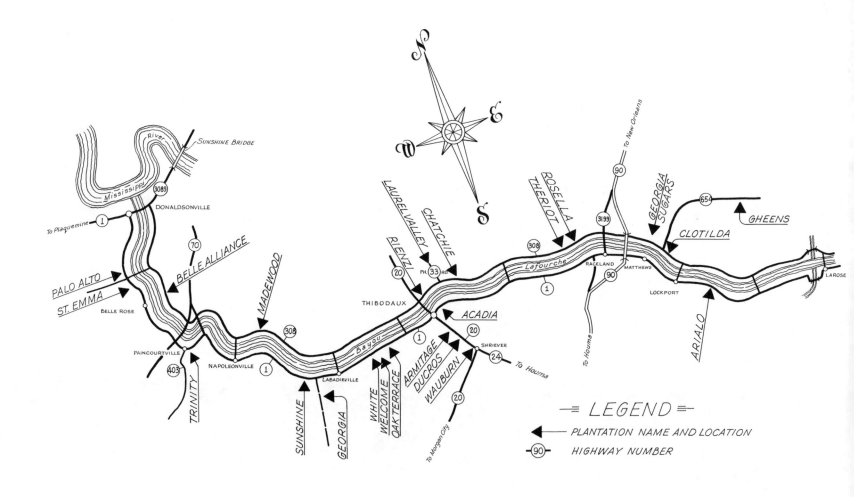

· LAFOURCHE PLANTATION HOMES ·

LAFOURCHE PLANTATION HOMES

EAST

Begin at the junction of Hwy 308 and Hwy 20 in Thibodaux. Travel 0.2 mile E on Hwy 308 to (L) *Rienzi Plantation Home** (1796). Louisiana colonial, good condition; private.

Continue 1.8 miles SE on Hwy 308, turn left on Parish Rd. 33, and travel 1.7 miles to *Laurel Valley Plantation Home** (1834). Large complex of cabins, barns, and the house under restoration; private.

Return to Hwy 308 and travel 1.3 miles E to *Chatchie Plantation Home** (1856). Recently restored Louisiana colonial; private.

Return to Thibodaux to Hwy 1 at the junction with Hwy 20. Travel E 1.5 miles to (R) *Acadia Plantation Home** (1842). Modified Creole cottage, good condition; private.

From the junction of Hwy 308 and Hwy 3199 in Raceland, take Hwy 308 W 1.8 miles to (R) *Rosella Plantation Home** (1814). Beautifully restored Louisiana colonial; private.

Continue W 0.9 mile to (R) *Theriot/Khi Oaks Plantation Home** (1890). Louisiana Creole cottage, excellent condition, beautiful grounds; private.

From the junction of Hwy 308 and the bayou bridge at Matthews, travel N on the hard surface road 0.3 mile to (L) *Georgia Sugars Plantation Home*. Louisiana classic, good condition; private.

From the junction of Hwy 308 and Hwy 654 S of Matthews, follow Hwy 654 NE 6.3 miles to (R) *Gheens Plantation Home** (1840). British-American colonial, good condition, beautiful grounds; private.

Return to Hwy 308 and travel E 0.4 mile to *Clotilda Plantation Home* (1900). Creole raised cottage, good condition, beautiful grounds; private.

From the Hwy 1 bridge at Lockport, travel E 2.7 miles to *Arialo Plantation Home** (1862). Creole raised cottage, good condition; private.

WEST

Begin at the intersection of Hwy 20 and Hwy 1 in Thibodaux. Go 4.4 miles W on Hwy 1 to (L) *Oak Terrace/Margurite Plantation Home* (1820). Louisiana colonial, good condition; private.

Continue 1.6 miles W to (L) *E. D. White Plantation Home** (1790). Creole raised cottage, good condition; open to public.

Approximately 0.1 mile E of E. D. White Plantation Home, take the side road at the brick post and go 0.5 mile SW to *Welcome Plantation Home* (1790). Small cottage of unusual plantation architecture. Good condition; private.

Return to White Plantation. Travel Hwy 1 W 4.7 miles to the gravel road, turn left, and drive 1.3 miles to *Georgia Plantation Home*. Creole raised cottage, good condition; private.

Return to Hwy 1 and travel W 0.6 mile to (L) *Sunshine Plantation Home*. Creole raised cottage, good condition; private.

Continue W on Hwy 1 to Napoleonville. Cross over Bayou Lafourche to Hwy 308 and turn E 2.0 miles to (L) *Madewood Plantation Home** (1840). Greek Revival, good condition; open to public.

Return to Hwy 1 at Napoleonville. On Hwy 1 travel W 5.2 miles to Paincourtville to (L) *Trinity Plantation Home*. Creole raised cottage, good condition; private.

NOTE: A nominal fee is required for most homes open to the public.

* *Home included in* HISTORIC BRIEFS.

At Belle Rose bridge, cross Bayou Lafourche to Hwy 308, and turn NW 0.5 mile to (R) *Belle Alliance Plantation Home** (1846). Large Louisiana classic, good condition; private.

Return to Hwy 1 at Belle Rose and go NW 2.6 miles to (L) *St. Emma Plantation Home**. Louisiana colonial, fair condition; private.

Go approximately 0.5 mile farther NW to the intersection of Hwy 1 and Hwy 940 to (L) *Palo Alto Plantation Home** (1850). Creole cottage, good condition; private.

SOUTH

Begin at the intersection of Hwy 20 and Hwy 3107 in Thibodaux. Go 0.9 mile S on Hwy 20 to (R) *Armitage Plantation Home* (1852). Creole raised cottage with square columns, good condition; private.

Continue 0.7 mile S on Hwy 20 to (R) *Ducros Plantation Home** (1833). Louisiana classic, good condition; private.

Proceed 0.5 mile S to the intersection with Hwy 24 to (R) *Wauburn Plantation Home* (1875). Creole raised cottage, good condition; private.

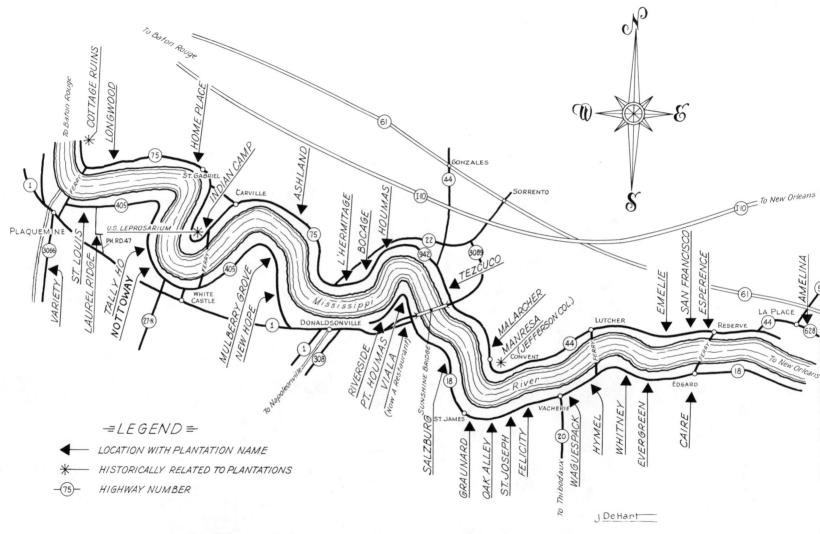

• OLD RIVER ROAD PLANTATION HOMES •

OLD RIVER ROAD PLANTATION HOMES

WEST BANK

Begin on Hwy 18, 27.5 miles W of the Huey Long Bridge (about 14 miles E of Vacherie) at *Caire Plantation Home*. Creole raised cottage with a complex of buildings, a store, another Creole cottage, and an old bank building from the 1880s. Good condition; private.

Continue 5.2 miles on Hwy 18 to *Evergreen Plantation Home** (1830). Louisiana classic, complete with practically all outbuildings: kitchen, guesthouse, *garçonnières*, dovecotes, stables, slave cabins, privies, and the overseer's house. Beautiful grounds with large oaks, magnolias, and old fences. House in good condition; private.

Proceed 1.0 mile on Hwy 18 to *Whitney Plantation Home and Store** (1840). Louisiana colonial with dovecotes and outbuildings. Fair condition; private.

Continue 4.0 miles on Hwy 18 to *Hymel Plantation Home*. Louisiana colonial, badly deteriorating; private. Another smaller home is located on the grounds about 100 yards W; it too is badly deteriorating.

Go on 1.6 miles to *Waguespack Plantation Home* (1825). Louisiana colonial with guest house (to far left). Good condition; private.

Ahead 2.5 miles is *Felicity Plantation Home** (1850). Large Louisiana colonial, good condition; private.

Continue 0.4 mile to *St. Joseph Plantation Home** (1820). Louisiana colonial with slave cabins and other outbuildings. Good condition; private.

Proceed 0.3 mile to *Oak Alley Plantation Home** (1836). Louisiana classic, excellent condition; open to the public.

Continue 4.8 miles to *Graunard/St. James Plantation Home*. Creole (West Indies) raised cottage with giant oaks and slave cabins. Good condition; private.

Go on 5.2 miles on Hwy 18 to *Salzburg Plantation Home*. Creole raised cottage, good condition; private.

Travel 5.1 miles farther to the access road to the Sunshine Bridge, Hwy 3089, to *Viala Plantation Home*. Now Lafitte's Restaurant, this Louisiana colonial home was moved from Viala plantation on River Road near Donaldsonville. Open to public.

Return to Hwy 18 below Sunshine Bridge, and proceed 2.8 miles on Hwy 18 to *Point Houmas Plantation Home* (1858). Creole raised cottage, modified with screen and lattice on the porch to a degree that has altered the original appearance. Private.

Continue 2.7 miles to the office of Triad Chemical, *Riverside Plantation Home*. Creole cottage; private.

From the intersection of LA 1 and LA 405 at Evan Hall, N of Donaldsonville, turn on LA 405 and travel 1.5 miles to *New Hope Plantation Home*. Raised Creole cottage, good condition; private.

Continue 5.1 miles to *Mulberry Grove Plantation Home* (1836). Two-and-one-half-story Louisiana colonial. Good condition; private.

From the intersection of Hwy 69 and LA 405 in White Castle, travel 1.6 miles W on LA 405 to *Nottoway Plantation Home** (1857). Greek Revival with Italianate influence. This beautifully restored home also houses a gift shop and restaurant. Open to public.

Continue W 1.3 miles to *Tallyho Plantation Home**. Two-story Louisiana colonial, good condition; private.

NOTE: A nominal fee is required for most homes open to the public.

Proceed 10.8 miles on LA 405 to *St. Louis Plantation Home** (1858). Louisiana classic, good condition; private.

Drive 1.0 mile farther W on LA 405 to the free ferry and cross the river. In Plaquemine, travel 3.1 miles W on Hwy 3066 from the junction with LA 1, along Bayou Plaquemine, to *Variety Plantation Home** (1828). Louisiana Creole cottage with separate kitchen on the right and guest house on the left. Good condition; private.

From the junction of Hwy 1 and Parish Rd. 47, 4 miles S of Plaquemine, travel 0.5 mile E on Parish Rd. 47 to (L) *Laurel Ridge Plantation Home* (1870). Victorian home with bay windows, turrets, and arches. Very poor condition; private.

EAST BANK

Drive 1.4 miles on the Bonnet Carre Spillway bed road from the E levee of the spillway (or 22.9 miles from the Huey Long Bridge via Hwy 48 and the spillway bed road) to LA 628 at the W spillway levee. Continue 5.3 miles on Hwy 628 to *Amelina Plantation Home* (1830). Large Creole cottage, good condition; private.

Drive 0.2 mile to the junction with Hwy 44 in Laplace. From this point, go 5.9 miles W on Hwy 44 to *Esperance Plantation Home*. Louisiana Creole raised cottage, West Indies design. Fair condition; private.

Continue 2.2 miles to *San Francisco Plantation Home** (1850). Gothic style, good condition, nice grounds; open to public.

Proceed 0.4 mile to *Emelie Plantation Home*. One-and-one-half-story Creole raised cottage with large belvedere. Good condition; private.

Go on 14.5 miles to *Manresa House of Retreats/Jefferson College** (1831). Large Greek Revival with impressive round columns across the front. Excellent condition; open by appointment.

Continue 4.3 miles to *Malarcher Plantation Home** (1891). Creole raised cottage, good condition; private.

Proceed 6.0 miles to Sunshine Bridge and go on 1.1 miles to *Tezcuco Plantation Home** (1855). Louisiana Creole cottage, good condition; open to public.

Continue 2.4 miles to *Houmas House** (1840). Louisiana classic, beautiful grounds. Good condition; open to public.

Travel on 1.9 miles on Hwy 74 to *Bocage Plantation Home** (1801). Greek Revival, good condition; open by appointment.

Continue 1.0 mile to *L'Hermitage Plantation Home** (1812). Louisiana classic, good condition; private.

Go on 7.4 miles to *Ashland/Belle Helene Plantation Home** (1841). Louisiana classic, needs repair.

Continue 7.0 miles to *Indian Camp/Woodlawn Plantation Home** (1857). Indian Camp Home now serves as the administration building for the U.S. Public Health Service Hospital at Carville. Louisiana classic, excellent condition; guided tours.

Drive on 8.1 miles to just N of the intersection of Hwy 75 and Hwy 74, at St. Gabriel, to *Home Place Plantation Home** (1790). Louisiana colonial, wooden columns and beautiful grounds. Beautifully renovated; private.

Continue 16.0 miles N on Hwy 75 to *Longwood Plantation Home* (1795). Louisiana classic, good condition; private.

Continue 5.8 miles farther to see the ruins of a plantation cottage; only the columns were left after a fire.

BATON ROUGE

At 8151 Highland Rd. is *Mount Hope Plantation Home** (1817). Creole raised cottage, beautiful condition; open by appointment (check with Baton Rouge tourist information).

Located at 2161 Nicholson Dr. is *Magnolia Mound Plantation Home** (1791). Creole raised cottage, good condition; open to public.

• NEW ORLEANS AREA PLANTATION HOMES •

NEW ORLEANS AREA PLANTATION HOMES

NEW ORLEANS

At 5023 Dryades St., on the corner of So-niat St., is *Soniat-Dufossat Plantation Home* (1840). Louisiana colonial, good condition; private.

On the U.S. Naval Base in Algiers is *Ott Plantation Home* (1830). Louisiana colonial, good condition; private.

Lombard Plantation Home (1826) is at 3933 Chartres St. Creole cottage, good condition; private.

Located at 305 Claiborne Ct., off Jefferson Hwy, is *Oaklawn Plantation Home*. Cottage with eight round columns, good condition; private.

At 3 Garden Lane is *Hurst-Williams Planta-tion Home* (1832). Louisiana colonial, good condition; private. The home was originally located on the corner of Tchoupitoulas St. and Nashville Ave.

Aurora Plantation Home (1742) is at Patter-son Rd. and Westchester Pl. in Algiers. Loui-siana colonial; private.

At 2340 Prytania St. is *Westfeldt Plantation*

Home (1830). Louisiana colonial, good condi-tion; private.

Several plantation homes may be found on Bayou St. John. At 924 Moss St. is the *Louis Blanc House* (1793). Louisiana colonial, good condition; private.

Plantation House/Spanish Custom House (1784) is at 1300 Moss St. Louisiana colonial, good condition; private.

At 1342 Moss St. is *Evarista Blanc Plantation Home* (1834). Louisiana colonial, good condi-tion. Private—now serves as the rectory for Our Lady of Rosary Catholic Church.

Down the block at 1347 Moss St. is *Wisner House* (1859). Raised Greek Revival cottage, good condition; private.

Located at 1440 Moss St. is *Pitot House* (1799). Louisiana colonial, good condition; open to public.

Musgrove Plantation Home (1859) is at 1454 Moss St. Louisiana classic, good condition; private.

EAST

Presently serving as the office of American

Sugar Refining Company, at 7417 N Peters in Arabi, is *Villere Plantation Home*. Louisiana classic, good condition; private.

On LA 39, located at Chalmette National Park, is *Beauregard House** (1840). Louisiana classic, formerly called Bueno Retiro. Open to public.

On LA 46, travel E 4.6 miles from Hwy 39 to (L) *Kenilworth Plantation Home** (1759). Louisiana colonial, good condition; private.

NORTH

Begin at Covington, Louisiana. From the junction of Hwy 190 and Hwy 21, travel 4.0 miles N on Hwy 21 to Hwy 21 and Oswald Rd. (R), the site of *Sunnybrook Plantation Home** (1880). Louisiana colonial, beautiful grounds. Very good condition; private.

From the junction of Hwy 190 and Hwy 21, travel 0.2 mile S on Hwy 190 to Riverside Dr. (R). Proceed 0.7 mile on Riverside Dr. to the entrance gate of *Villa de la Vergne** (1820). Large French villa style plantation home with bell tower to one side. Fair condition; private.

NOTE: A nominal fee is required for most homes open to the public.

* Home included in HISTORIC BRIEFS. 95

At Beau Chene Estates on Hwy 22, E of Madisonville, is *Beau Chene Plantation Home*. Louisiana colonial, good condition; private.

In Amite, Louisiana, E of U.S. 51 at the end of Elm St., is *Blythewood Plantation Home*. Modified Greek Revival with surrounding lower galleries and central columns extending from the lower floor to the upper balcony galleries. Fair condition; private.

WEST

To see homes W of New Orleans on the West Bank of the river, begin on LA 18 at Huey Long Bridge. Travel E 1.6 miles from the bridge to *Magnolia Lane Plantation Home*. Louisiana colonial, needs repair; open to public.

Continue E 0.1 mile to (R) *Derbigny Plantation Home* (1826). Creole raised cottage, presently a riding academy. Good condition; open by appointment.

Return to Huey Long Bridge and travel W 4.2 miles to (L) *Tchoupitoulas Plantation Home* (1850). Modified Louisiana colonial, now a restaurant. Open to public.

Continue 14.3 miles to (L) *Fashion Plantation Home*. Louisiana colonial, good condition; private.

Proceed 1.0 mile to (L) *Home Place/Keller Plantation Home** (1790). Louisiana colonial, good condition; private.

Continue 8.0 miles to (L) *Glendale Plantation Home** (1790). Louisiana colonial, good condition; private.

For homes W of New Orleans on the East Bank, begin on LA 48 at Huey Long Bridge. Go E 0.5 mile from the bridge on River Rd. (L) to *Whitehall Plantation Home* (1852). Louisiana classic now serving as Magnolia School. Good condition; private.

Return to Huey Long Bridge and travel W 0.4 mile on Hwy 48 to (R) *Elmwood Plantation Home** (1762). This home was being used as a restaurant, but was practically destroyed by fire in December 1978. Plans are being made to renovate. Modified Louisiana colonial; private.

Continue 10.4 miles to the junction with Hwy 50 at St. Rose. Travel 4.5 miles farther on Hwy 48 to (R) *Destrehan Plantation Home** (1787). This Louisiana classic is one of the better examples of Louisiana antebellum homes. Open to public.

Continue 1.4 miles on Hwy 48 to (R) *Ormond Plantation Home** (1790). Louisiana colonial, good condition; private.

WHITE PILLARS AND SPIRITUAL SPIRES

J DeHart

Spiritual Spires

White Pillars and Spiritual Spires _____

As we travel the shoulders of the historic Mississippi in search of remnants of early plantation life, we find many sites with evidence of a once-thriving planter's domain. We see many clusters of great oaks that once sheltered the manors, waiting once again to be of such need. We find mills, barns, cabins, and pillared mansions still alive. We look among the two-hundred-year-old trees for fragments of Welham, Hester, Belmont, and others on the east bank of the old river. But in our haste to discover much we often find little. We strive to capture it all within the scope of our imagination and we often lose very much.

But just above the town of Lutcher, near the little village of Convent, we cannot overlook one of the most astounding sights of these old byways. Suddenly, without warning, bursting forth from the very earth, we come upon twenty-two magnificent white columns all amassed before a huge ancient monument! It is as though all the stately mansions of this old river had gathered together in solemn convention. And behind those pillars stands a dazzling white building with splendid proportions in the Greek Revival spirit. All this is set far back on a wide green meadow eloquently framed by antique oaks, camphors, and magnolias.

Here is Jefferson College! This imposing structure is a symbol of the high learning and culture that existed a century and more ago. It is now buried far away where few can behold its splendor; this massive shrine to dignity and the high values of man. This great jewel is now a religious retreat called Manresa and is under the direction of Jesuit priests. It was built in 1830 as nonsectarian institution of higher learning for the sons of Louisiana planters. During the golden era of plantations this was the king of colleges, where young men were molded for future responsibilities and were taught to cherish and preserve the heritage that had built this great country. It was a focal point for culture in harmonious tranquility, when life was guided on a course of dignity and self-respect.

During the Civil War Jefferson College as taken over by federal troops to be used as a barracks. In 1864, after having been purchased by Valcour Aime, the man widely known as the "prince of

Louisiana planters," it was donated to the Roman Catholic Marist Fathers to be turned once again into an institution of higher culture. In 1931 the Jesuit priests took it over and renamed it Manresa House. It remains today a retreat for the layman and is as imposing as it ever was. At the left of this large structure and nearer the road stands a two-story manor that reflects the pure Louisiana Classic that has been used as a home for priests and educators for more than a century.

One can imagine the awe-strickened riverboat passengers as they may have looked through the avenue of oaks to view those twenty-two magnificent columns that front the building that so well symbolizes a great gift to mankind. For nearly three-quarters of a century the Mississippi River ruled the nation. Such dominant factors as plantations and such imposing monuments as Jefferson College played a very important role in the lives of all.

But the true gem of Manresa acres is a quiet miniature not easily viewed from the roadway. In great contrast to the massive pillars of overpowering prominence is the graceful little chapel of Gothic design that is set away in the right corner of the meadow beneath protective live oaks. Its tasteful spires reach up into the massive branches, dripping with Spanish moss, entirely encompassing the little treasure in shadows and making it all but obscure from the passerby. No other church can better reflect the solitude, the character, and the atmosphere of the old South. In no other place can man find more peace and spiritual enhancement than this chapel, where reverence is so well preserved. If any sound can be heard, it is one's own heartbeat, quickened by the excitements of sacred tranquility.

Here it is easy for a person to reflect on images of the pasts of those who helped to mold that golden age. Envision if you wish the tall, slender, and gracious Valcour Aime strolling gently along the walkway, speaking anxiously with the college dean about future plans for the school. It was Valcour Aime who built the beautiful little Gothic chapel in loving memory of one of his daughters.

If a dignified culture must be guided by a higher spirit, then this atmosphere lends itself well for inspiration. If tradition can be wholly revived into long-established meaningful customs of merit, culture, and Christian practices, then methods to make these effectively flourish again could be well stimulated within the tranquil shadows of this wonderful little chapel. This is a place where one can truly listen to the word of the Lord.

HISTORIC BRIEFS
Prominent Old Homes

Prominent Old Homes

Acadia (Lafourche)

A short distance below Thibodaux on Bayou Lafourche, Acadia almost invites the traveler to its refreshing shaded grounds. The rambling one-story cottage was built in 1842 by Phillip Key, a relative of Francis Scott Key.

The present house is built on the site of a plantation established by Jim Bowie and his brother Stephan, who once had a very lucrative slave-buying business with Jean Lafitte. They built the plantation on this location at the suggestion of Lafitte.

Acadian House (Attakapas)

Located on Bayou Teche in the beautifully wooded Longfellow-Evangeline State Park, Acadian House is a showpiece of Louisiana history. There are many samples of furnishings, clothing, and tools of early Louisiana in this home, which serves as a museum.

It was erected in about 1765 for Chevalier D'Hauterive, servant of the King of France, and commandant of the Poste des Attakapas (now the town of St. Martinville). Poste des Attakapas was established for the protection of the early Acadian exiles from Nova Scotia who began to settle in Louisiana during the 1750s.

The old home is constructed from hand-hewn and pegged cypress logs, handmade sunbaked bricks, and bousillage walls. There is an outdoor kitchen furnished with antique utensils and a smokehouse in the rear of the home.

The ancient and magnificent "Gabriel Oak" graces the front lawn of Acadian House. The old home, which is listed on the National Register of Historic Places, is open to the public for a modest fee.

Alice/Fuselier

Albania (Techeland)

The Grevemberg family, who as supporters of the crown fled from France during the revolution, built this large Louisiana classic mansion in 1842. A grove of huge live oaks shades the peaceful grounds that slope gradually to the banks of old Bayou Teche.

In 1854 a Jamaican named Isaac Delgado bought the plantation and became in a short while a very wealthy sugar planter. In 1910 he bequeathed the plantation, including the mansion and huge sugar mill, to the city of New Orleans so that its profits could be used to maintain the Delgado Trade School and the Delgado Museum of Art, both of which he had also donated to New Orleans. The city operated the plantation until 1957, when it was sold to private owners.

Albania is furnished with antique pieces and is open to the public for a reasonable fee.

Alice (Techeland)

Originally known as the Fuselier House, Alice was built about 1800 by Agricole Fuselier near the present town of Baldwin. In later years the home was transported by barge to the present location on Bayou Teche. Standing today as one of the oldest homes in the Teche country, Alice is privately owned.

Allendale (Dixie-Overland)

Colonel John James Marshall came to north Louisiana in 1854 and built a one-room squatter's log cabin. Additions were made as time went on, and the building eventually housed five generations of Marshalls. The original log cabin remains as part of the existing home.

In 1886 the north Louisiana dairy industry began at Allendale with the first registered cattle. The Marshall family also built the All Saints Episcopal Church, which is located off Frierson Road just west of Linwood Road. Beautifully furnished and preserved, the little church is one of the last remaining antebellum churches in Louisiana. Although the plantation and home are private, the little church can be visited by making arrangements with the owners of Allendale. Check with the local tourist information center for further information.

Ardoyne (Terrebonne)

Located north of Houma, Ardoyne, a Victorian Gothic mansion, was built of virgin cypress in the 1890s by John D. Shaffer. Shaffer operated his own plantation railway with a locomotive and 160 cane cars to transport crops to the sugar house. The impressive mansion has been carefully maintained in recent years by the Lee Shaffer, Sr., family. Ardoyne is a private residence that is open by appointment only.

Arialo (Lafourche)

Arialo is said to have been constructed by Joseph Claudet in 1862, during the occupation of south Louisiana by Union forces in the Civil War, and is perhaps the only plantation home built during that time period. Joseph Claudet purchased the land, which was originally acquired in 1780 by Jacque Lamotte in a Spanish land grant, from Augustin Cunio. Recently renovated, it is now a private residence; it can be easily viewed from the highway just below Lockport.

Arlington (Dixie-Overland)

Arlington, a two-story classic home located on Lake Providence, was an important Union Army headquarters during the Civil War. Fortunately, it was saved from destruction, in contrast to the many beautiful plantation homes that were destroyed by fire at the hands of federal troops.

Edward Sparrow, a native of Ireland, bought the home in 1852. It was originally a one-story house that was built in 1841; Sparrow lifted the cypress dwelling and constructed a new brick floor underneath it. General Edward Sparrow became senior senator from Louisiana in the Congress of the Confederacy.

Across the lake from the town of Lake Providence, Arlington nestles in a grove of ancient magnolias, live oaks, cypresses, cedars, and black walnuts. Now a private home, Arlington is an impressive view from Highway 65.

Asphodel

Arlington (Techeland)

Near Franklin, Arlington, a well-preserved Greek Revival mansion, was built with slave labor in the 1850s by Euphrazie Carlin. Carlin, a very wealthy mulatto, had hundreds of slaves to operate his vast holdings.

Arlington is now a private residence that can be enjoyed by travelers on the old "Spanish Trail."

Ashland/Belle Helene (Old River Road)

On the old River Road above the Sunshine Bridge is the Ashland Plantation Home, built in 1841 by Duncan Kenner. Kenner was a very wealthy planter who served at one time as the Confederate minister to France and later as U.S. tariff commissioner under President Arthur. At the age of seventy-one, he became chairman of the Cotton Centennial Exposition in New Orleans.

The home was renamed Belle Helene when it was purchased by John B. Reuss. It developed into one of Louisiana's great sugar plantations.

Twenty-eight large square columns completely surround the two story mansion, which features wide galleries and an interesting facade. With its unusual height, it nestles in a setting of huge trees a short distance from the old Mississippi River, which brought many visitors during the flourishing sugar age.

Belle Helene was used in 1957 for the filming of the movie "Band of Angels" starring Clark Gable; but the old home has recently fallen into decay. The Louisiana State Department of Recreation and Tourism has awarded a grant to help repair Ashland, which the Shell Oil Company has matched. The building is open to the public.

Asphodel (Felicianas)

Naming his home after the daffodil, Benjamin Kendrick built Asphodel in 1835. It soon became the center of a prosperous cotton plantation.

The large cottage, with a dormered gable roof supported by six large round columns, has a recessed wing on both sides. The old home has seen a number of famous artists and writers who came to work, visit, or stay for a while. Among the better known are the artist John James Audubon and author Lyle Saxon, who is famous for his novels on old Louisiana.

Austerlitz (False River)

A Louisiana colonial home erected in 1832, Austerlitz was named for one of Napoleon Bonaparte's major battles. A West Indies architect designed the home, which was later enlarged to bring the total number of rooms to twenty-two. Large fourteen-foot-deep galleries completely encircle the structure. The spacious gardens are bordered by large pecan trees that were planted at the time the house was built.

Bayou Folk Museum (Red River)

At Cloutierville, on the beautiful Cane River Lake, is an authentic restoration of the home of authoress Kate Chopin. The Louisiana colonial home is built of hand-hewn cypress, sunbaked bricks, and bousillage, and has been converted into a museum that houses many Cane River country relics. Included in the display are a blacksmith shop, a country doctor's office, and many early farm implements. The museum is on the National Register of Historic Places and is open to the public for a small fee.

Bayside (Techeland)

Located on Bayou Teche near Jeanerette, Bayside was built in 1850 by Francis D. Richardson, a member of the pre–Civil War Louisiana legislature and a classmate and close friend of Edgar Allan Poe. Bayside, a private home, is a beautiful Louisiana classic structure resting in the tranquil setting of the historic Teche country.

Beau Fort (Red River)

Built for Narcisse Prudhomme I in 1790, the sprawling single-story cottage at Beau Fort has an eighty-four-foot-wide front gallery that still has the original flooring. Also featured are the huge underground cisterns that once served Fort Charles. The home, now private, is listed on the National Register of Historic Places.

Beauregard

Beau Pre (Techeland)

A beautiful old home on the tranquil Teche, Beau Pre was erected in 1830 by John Jeanerette, the founder of the nearby town bearing his name. Legend states that Longfellow was a guest here at one time and that it was here that he was inspired to write *Evangeline*.

Beauregard House (New Orleans)

Located in Chalmette National Park, east of New Orleans, Beauregard House was built in 1840 on the site of the Battle of New Orleans. The nearby Macarty plantation home was where Gen. Andrew Jackson made his headquarters in 1814 and 1815 during the anxious weeks that New Orleans was threatened by the British.

Beauregard House, also known as the Bueno Retiro plantation home, was later owned by Judge Rene R. Beauregard, son of the Creole general of Civil War fame. It was restored in 1958 by the federal government and is the visitors' center for Chalmette National Park. Overlooking the grand battlefield and the Chalmette National Cemetery, the historic site is open to the public.

Belle Alliance (Lafourche)

Belle Alliance, an elaborate Creole classic home, was built in 1846 for the German consul general to New Orleans, Charles Koch. Large square columns cross two level galleries with iron grillwork and an outsweeping stair dominate the thirty-three room manor. Koch, who was originally a Belgian aristocrat, used Belle Alliance as his family home until 1915. He also built St. Emma, which stands today a short distance across Bayou Lafourche.

The Belle Alliance sugar mill, which processed the plantation's 7,000 acres of sugar cane each year, was considered to be the most important sugar mill west of the Mississippi River. Although the mill no longer exists, remnants of other supporting structures and slave cabins can still be seen nearby.

Belle Alliance

J DeHart

Bocage (Old River Road)

Bocage, whose name means "shady retreat" in French, was built in 1801 by Marius Pons Bringier as a wedding present for his fifteen-year-old daughter on the occasion of her marriage to Christophe Colomb, a Parisian. Bocage is an impressive plantation home built on the Louisiana classic design. Privately owned, Bocage is open by appointment.

Bocage (Techeland)

A large Greek Revival structure, Bocage was constructed around 1846 on a plantation known as Oakbluff, seven miles away from its present location. It was transferred to the present site by barge on Bayou Teche, which was an outstanding accomplishment for such a large structure. In a setting of beautiful old live oaks, Bocage makes a striking impression to travelers of the old Spanish Trail.

Burns (Tensas)

Zenith Preston built the sprawling one-and-one-half story cottage known as Burns in 1853. The large rooms of the home are covered with a wide multiple dormered hip roof that has interior chimneys, and the building is surrounded by 300 feet of galleries. The large rooms are plastered and frescoed. Handsome marble mantels frame the fireplaces. The doors are trimmed with silver.

Callihan (Red River)

Callihan is a large Creole cottage with a gabled roof, three dormers, flanking chimneys, and six square columns across a spacious gallery. As with most old plantation homes, the original kitchen, located behind the main building, was never a part of the house.

The home was built in 1841 by James Callihan, who operated the plantation for over thirty years. It was a large cotton plantation, but some sugar cane was also raised. After changing hands several times, it is now the private residence of Mr. and Mrs. H. S. Coco, who acquired it in 1967.

The house originally faced Bayou des Glaises; in 1927 it was turned around to face the highway.

J. DeHart

Chrétien Point

Cedar Grove (Red River)

Cedar Grove has high square brick columns rising to a second level gallery which has shorter square wooden columns joined together with ornate balustrades. Although the house is of the design prevalent in the early 1820s, there is some evidence that it was first constructed in the mid 1700s and subsequently altered in the early nineteenth century.

Chatchie (Lafourche)

A two-story home with hip roof and very large dormers, Chatchie was built in the 1850s. It was in very poor condition until recently, when the present owners made extensive repairs and renovated it into one of the most beautiful homes now existing on Bayou Lafourche. A private residence, Chatchie can nevertheless be viewed and appreciated from the highway a few miles south of Thibodaux.

Chretien Point (Attakapas)

Five hundred slaves once worked at Chretien Point plantation. The old home was built in 1830 by Hypolite Chretien at the same time that the Shadows was built in New Iberia by David Weeks; some believe that the same architect was used for both structures.

Chretien Point is a regal manor with a great history and stories that reach back to the pre–Civil War days. The plantation had a gambling, cigarette-smoking mistress who was a match for any man with a pistol. Privateers Jean and Pierre Lafitte were frequent visitors here.

Chretien Point is open to the public by appointment only.

Cottage (Felicianas)

A low rambling one-and-one-half story house with an exceptionally long gallery and dormered roof, Cottage is flanked by a number of buildings that were erected during a period from 1795 to 1859. The main house was itself built in 1811 completely of virgin cypress.

In 1815, returning from the Battle of New Orleans, Gen. Andrew Jackson and his staff were guests at Cottage. The house contains many historic relics and letters from great figures of American history.

The Cottage is open for a fee and has lodging facilities for the overnight traveler.

Crescent (Dixie-Overland)

Crescent narrowly escaped destruction from Grant's soldiers during the Civil War. A sympathetic young officer spared the house from fire because of the gravely ill mistress that lay in a second story bedroom.

This two story classic manor still has the brass doorknobs and locks that were installed when it was built in 1832. The massive front door is flanked by French windows fourteen feet tall. The spiral staircase and the stained glass transoms imported from Europe remain as beautiful today as they were when installed.

Destrēhan

Crescent Farms (Terrebonne)

Crescent Farms is the second oldest home standing in the area of Little Bayou Black above Houma. Built in 1834 by William Shaffer, it has been beautifully restored in recent years. Although a private residence, it is easily viewed from the highway.

Darby (Techeland)

Built around 1765, Darby now houses a bank in the little community of Baldwin on the banks of the famous Bayou Teche. François Darby was an early owner of the home; it was later bought by John Baldwin, the man the town was named for.

Destrehan (New Orleans)

Destrehan is located on the east bank of the old River Road just a few miles above New Orleans. Built in 1787 for Robert Antoine Robin de Logny, it was purchased in 1802 by his son-in-law Jean Noel d'Estrehan de Beaupre, the son of the royal treasurer of the French colony. Destrehan was a very large plantation at that time.

There are many stories of frequent visits by the ghosts of Jean Lafitte, Governor Claiborne, and other notable spirits to the old rooms of Destrehan. The wife of Governor Claiborne was a frequent visitor to the plantation. Rumors that Destrehan was the secret rendezvous for Jean Lafitte and a prominent married lady of the elite society of New Orleans became spicy gossip during the early 1800s.

The old home was stripped by thieves as time went on and eventually fell into a very sad state of neglect and abuse. In 1972 the American Oil Company, which owned the property, donated the house and four acres to the River Road Historical Society. The Society has made giant strides toward the restoration and preservation of this beautiful Louisiana treasure and now offers it for the public to enjoy for a modest fee. Some outbuildings of Louisiana Creole design have been moved near the main manor as a gift shop and tea room.

Dixie (Techeland)

A two-story home, Dixie was built in 1850 by the Richard Wilkins family. It was later purchased by Murphy J. Foster, who became Louisiana's thirty-first governor in the 1880s and U.S. senator in the early 1900s. Dixie has been privately owned by his descendants ever since.

Dogwood (Felicianas)

Originally a log cabin that is now enclosed in the present home, Dogwood is a beautiful Creole cottage with a dormered hip roof. Located southeast of St. Francisville, it is a private home.

Ducros (Lafourche)

Built in 1833, Ducros is said to have been modeled after Hermitage. The original owner, French doctor M. Ducros, later sold it to Van F. Winder. With a total acreage of 5,600, it was the first large sugar plantation established in Terrebonne Parish.

Edgewood (Dixie-Overland)

The captivating Edgewood, also known as the Baughman plantation home, was constructed in 1900. Two attractive entrance stairs front the sprawling gallery. Another outstanding feature of the Victorian design is the turret tower at the center of the structure. A little over a mile from Farmerville, Edgewood is a private home that can be easily viewed from the roadway.

Ellerslie (Felicianas)

Originally a cotton and cane plantation, Ellerslie was built in 1832, according to legend as a rival to the huge Greenwood manor house just a few miles away. It was an architectural duel between two plantation owners, Judge William Wade of Ellerslie and Ruffin G. Barrow of Greenwood. The duel went on for four years, with some $200,000 involved. Nearby forests and brick kilns furnished most of the materials for Ellerslie and slaves furnished the labor. Wade lived in his mansion for fifteen years and was buried in the cemetery on the rear grounds.

Ellerslie eventually won over Greenwood by default since the latter burned to the ground. Ellerslie still stands as the great manor of a large present-day cattle ranch.

Elmwood (New Orleans)

Elmwood was built in 1762 by Nicholas Chauvin de la Freniere on a land grant dating back to 1719. It is perhaps the oldest plantation home in the South. The first American governor of Louisiana, W. C. C. Claiborne, once lived at Elmwood.

Originally the old house was two and one-half stories high, but in 1940 a fire devastated the upper levels. It was restored as a single-story mansion and was later turned into a fine restaurant.

In December 1978 Elmwood was taken by fire, once again destroying a good portion of the building. Plans are reportedly being made to restore the home into a restaurant.

The old building is in a setting of beautiful grounds and massive ancient live oaks shrouded with Spanish moss.

Enterprise (Techeland)

An attractive home that still serves as the "big house" of a present-day sugar plantation, Enterprise was built on classic lines in the 1850s. The plantation was established by the Patout family for the purpose of raising grapes for the wine industry. With soil unsuited for growing grapes, the sugar crop was established; descendants of the Patout family still operate the plantation and its huge sugar mill.

Evergreen

Evergreen (Old River Road)

On the west bank of the Mississippi near Edgard, Evergreen was built around 1830 by Michel Becnel and in 1840 was purchased by Ralph Brou, a wealthy planter. Plans and dimensions of this River Road gem have been stored in the federal archives in Washington.

Evergreen has been beautifully restored by its present owner, Mrs. Harold Stream of New Orleans, who has preserved not only the mansion but also the many outbuildings. There are *garçonnières*, *pigeonniers*, servants' quarters, barns, slave cabins, outdoor privies in Greek Revival design, and a beautifully restored overseer's home, which was the main house before the mansion was built. The entire complex is on private property, but much of it can be viewed from the west bank River Road.

Faye (Techeland)

Located on Lewis Street in Jeanerette is Faye, perhaps one of the best examples of a Louisiana Creole cottage built during the mid 1800s. Old Faye was moved to Lewis Street from the Faye plantation, which was located two miles east of Jeanerette. Very tastefully preserved, this private residence can be viewed easily from Lewis Street.

Felicity (Old River Road)

Felicity was built by the wealthy and famous Louisiana planter Valcour Aime in 1850 as a wedding present for his beautiful daughter Felicite. It boasts of red Italian marble mantels over the fireplaces. Located on the old River Road west of Vacherie, it is now a private residence.

Ferry (Tensas)

Near Sicily Island in Catahoula Parish, Ferry overlooks the beautiful Lovelace Lake, which was a popular hunting ground for the Tensas, Chickasaw, Choctaw, and Natchez Indians. The old two-story home, built in 1830 of pine and cypress, essentially retains its original appearance today.

There are several Indian mounds on the grounds; many artifacts of Indian battles have been found on the site. It is generally agreed by historians that the present location of Ferry was where the Natchez Indian nation made its last stand against the French in the latter part of 1730. The French were provoked to the attack by the horrible massacre of 1,500 men, women, and children at Fort Rosalie in 1729.

Still an operating cotton plantation, Ferry features many remaining supporting buildings of its early days.

Filhoil/Logtown (Dixie-Overland)

A one-story cottage, Filhoil is joined together with pegs. The rooms have no ceilings, and the large exposed rafters have been carefully finished and painted. The dining room was designed to resemble the dining room of an old river packet steamer. All hinges and locks were made by the plantation blacksmith.

Filhoil was built in 1855 by John B. Filhoil, a wealthy planter, and his grandson Don Juan Filhoil, the commandant of the Poste des Washita, now the city of Monroe. Located on the east bank of the Ouachita River twelve miles south of Monroe, Filhoil is still a private home.

Frithland

Forest (Felicianas)

Forest plantation, a cotton and corn plantation, was originally owned by Augustine Allain. The first home built on the land was owned by Mr. and Mrs. Alexander Powell, but it was burned to the ground by Union soldiers in 1863. The Powells died shortly afterwards, within a year of each other, and were buried in Grace Church cemetery at St. Francisville.

The existing home, built in 1894 by the McQueen family, is presently owned by Mr. and Mrs. E. R. Broadbent. The Broadbents bought Forest in 1969 and have extensively restored the sprawling cottage, which can be enjoyed from the highway.

Frances (Techeland)

Built in 1820 by Louis de Maret, Frances was owned by his family until 1876. It was then purchased by Louis Kramer, who gave it the present name in honor of his daughter.

Now open to visitors, Frances houses a gift shop and displays of antique pieces. On the beautiful grounds there are several other supporting buildings and structures of early plantation life.

Frithland (Red River)

Frithland is a beautiful classic plantation home at the southern limits of the town of Bunkie. The delicate, almost feminine, ornate metal railing and balcony and the gracefully fanlighted entranceway reflect a queenly presence. The magnificent fluted columns, featuring powerfully dressed capitals, and the elegant side doorways on the other hand seem to be of a masculine dignity befitting a well-bred prince.

Gheens (Lafourche)

Still the center of a complete plantation, Gheens is now the "big house" for the large Golden Ranch. The antebellum home was restored in recent years by Koch and Wilson Architects.

Near the house are the remains of a once productive sugar house. On the side road to the old mill stands one of the original brick slave cabins, which features large strap hinges on the windows.

Girard (Attakapas)

A Louisiana classic plantation home built in 1820, Girard is now the Old Acadian Inn in Lafayette. It has been beautifully restored as a very fine restaurant and is open to the public.

Glendale (New Orleans)

On the west bank River Road, Glendale is a treasured gem of early Louisiana architecture featuring brick and cypress construction. The large timbers of the house are hand hewn and fastened together with wooden pegs. All hardware is original; it was made by a slave blacksmith on the plantation. Two large underground cisterns and one dovecote are still in existence on the grounds.

Highland (Felicianas)

Highland was the first plantation home of Olivia Ruffin Barrow. Of a design popular in the early Carolinas and Tennessee, the home was built in 1805. It is a raised two-story frame structure built of virgin cypress, with six tall, square, wooden columns fronting long galleries.

Holleman/Segura (Techeland)

The present Holleman Louisiana colonial home was built using materials from the Segura Plantation home after it was nearly destroyed by years of vacancy and the damage of hurricanes. The T. C. Holleman family resurrected the beautiful home in the 1960s and it stands today as a sparkling example of true pride in the heritage of old Louisiana. Originally built in 1812 by Raphael Segura, this completely reclaimed home, situated amid moss-draped live oaks, now rightfully bears the name of the family who restored it to its original beauty.

Home Place (Old River Road)

A Louisiana colonial home, Home Place was built in 1790 by ancestors of the present owner, Simon LeBlanc. Mr. LeBlanc has reconditioned the old home into a very fine private residence.

Home Place/Keller (New Orleans)

On the west bank River Road near Hahnville, Home Place/Keller is of early Louisiana colonial construction. Built in 1790 by the Fortier family, it is an architectural treasure that has attracted many curious sightseers. The construction is very much the same as that of Parlange, which is located on False River.

Hopkins (Dixie-Overland)

Hopkins is a beautiful two-story home located in the town of Marion in north Louisiana. The serene atmosphere of the home, built in 1845, was the setting for a bit of musical history. In 1854, while serving as a music teacher for the children of homeowner Elias George, Ann Porter Harrison composed the beautiful song "In the Gloaming."

Houmas House (Old River Road)

Houmas House was built in 1840 by Col. John Smith Preston of South Carolina. It was purchased in 1857 by John Burnside for $750,000, a price which included the mansion and ten thousand acres of rich sugarcane land. Burnside was an orphan who worked to become Louisiana's wealthiest sugar planter. Burnside began to make his fortunes as a small business man in New Orleans. He also had the foresight and know-how to become the most successful bachelor of his time in Louisiana. Houmas plantation was only one of twelve plantations owned and operated by Burnside, who had over three thousand slaves.

Named for the Houmas Indians of the area, Houmas House was extensively restored in 1940 by Dr. George B. Crozat and is presently owned by his heirs. The home, elaborately furnished with antiques and placed in a setting of beautiful oaks, magnolias, and gardens, is open to the public for a fee.

Houmas House

Idlewild (Techeland)

On the banks of ancient Bayou Teche, Idlewild was built in 1850 by George Haydel. The large Creole cottage was restored by his grandson, E. H. Seyburn, in 1964.

Indian Camp/Woodlawn (Old River Road)

Located on the east bank River Road at Carville, Indian Camp Home is now the administration building for the Public Health Service National Hansen's Disease Center, the only hospital of its kind in the United States. The location of the medical center was once a part of the Indian Camp plantation (also known as Woodlawn), which was built in 1857 on the site of a Houma Indian village. There are strong indications in design and material suggesting that the famous architect Henry Howard, the architect of Nottaway and Madewood, also designed and built this manor.

The home, slave cabins, and outbuildings were turned over in 1896 to the Daughters of Charity of St. Vincent de Paul, a Roman Catholic order of nuns who assumed responsibility for caring for the Hansen's disease patients of Louisiana. It was converted into a federal hospital in 1921, but the Sisters of Charity still have an important role in the hospital affairs and operations.

The modern institution is composed of approximately one hundred buildings located on over three hundred acres of land. Almost a small town in its own right, the center has served Hansen's disease patients of the U.S. for over eighty-five years. Guided tours of the center and the buildings are conducted by staff personnel.

Ingleside (Dixie-Overland)

A large, unusual home with numerous small grouped columns, Ingleside has three stories, the first of plastered brick and the other two of wood. The home is reported to be haunted; it is said that at night chains clank in the attic, plantation bells toll, chairs rock, and melodies come from the ancient piano.

J DeHart

Indian Camp

Ingleside

Kenilworth

Kenilworth (New Orleans)

Kenilworth is a beautiful home located in St. Bernard Parish, east of New Orleans, that features many stories of ghosts. The most popular is of a headless man and woman who stroll through the rooms and up stairways each time the moon is full.

Kenilworth was built in 1759 as a combination residence and fortress by Pierre Antoine Bienvenue, who came to Louisiana in 1725 from Quebec, Canada. He became one of the richest and most influential men in the colony.

A solid structure with eighteen-inch walls, stout cypress doors, and massive wrought iron bolts and hinges, it was constructed entirely with wooden pegs, using no nails whatsoever. It was originally only one story high; the second floor was added in the early 1800s. Large wide galleries cool the beautiful Louisiana colonial home.

Kent (Red River)

Kent was built around 1800 by Pierre Baillo, the son of a French officer stationed at Fort St. Jean Baptiste (now Natchitoches). It is a fine example of the early Louisiana colonial home that resembled the West Indies planter's house, featuring bousillage construction with brick below and cypress above.

In 1842 the house was sold to Robert Hynson of Maryland, who added the two flanking bedroom wings. Hynson named the home Kent after his family's home in Kent County, Maryland.

The old home houses many early plantation furnishings, utensils, and other treasures. Located on four acres of land in Alexandria, Kent also has several supporting outbuildings of the pre–Civil War era. It has been beautifully restored and is open to the public for a fee.

Keystone Oaks (Attakapas)

Of Louisiana classic design, this house was originally built in 1870 as the overseer's home for the Keystone plantation near St. Martinville. The home was relocated at the end of the beautiful avenue of oaks after the "big house" burned in 1914.

Kent

J DeHart

Labatut (False River)

On the old west bank River Road across from St. Francisville, Labatut is an ancient relic of Louisiana history. Everiste Bana, a settler of Spanish descent, built this home in the late 1700s. The river has taken all of the oaks that once fronted Labatut and has encroached almost to the gallery of the home. Sadly neglected by the present owners, Labatut is private.

Labbe (Attakapas)

The old Labbe home, located near Keystone Oaks, is a large cottage with broad gallery and quite large dormers on the high gabled roof. It was once of Louisiana colonial design, but the lower floor was removed when the bricks deteriorated and the upper floor lowered, giving it its existing appearance. The home was built in 1850 by Eugene de Chastagnier as a wedding present for his daughter's marriage into the Labbe family. A son of that union later became a state senator.

Lakewood (Tensas)

On Lake Bruin near St. Joseph, Lakewood was built in 1854 by Capt. A. C. Watson, the commander of Watson's Battery during the Civil War. When leaving his home in 1861 to join Lee, Watson withdrew his entire fortune of $80,000 from the banks. He used $60,000 of it to buy equipment and furnishings for his regiment. The remaining $20,000 was buried on the grounds of his home; most of it was recovered when he returned from the war. However, in 1928 a descendant of Captain Watson, working a garden plot near the home, uncovered a jar containing $5,000 of the buried money.

Land's End (Dixie-Overland)

The architect for Land's End was M. Robbins, who built the home for Col. Henry Marshall in 1857. On ten thousand acres of rich Red River bottom lands in the hills of northern Louisiana, Land's End was a prosperous cotton plantation worked by three hundred slaves.

When Marshall's wife first came to the plantation from South Carolina, she exclaimed that she had been "carried to the end of the earth." The plantation's name consequently became "Land's End."

The house has many pieces of period furniture, books, and documents of historic value and is open to visitors by appointment. Check with the Shreveport tourist center.

Lane (Felicianas)

An 1825 plantation home, Lane has been very well cared for over the years. It was originally a two-story home with a single lower gallery and narrow colonnettes. In later years a kitchen was added on the west side and a bedroom on the east lower level. Excepting those additions, Lane remains today essentially as it was when it was built. It lies far back from the highway on a gravel road deep within a beautiful dense forest southwest of Clinton.

A descendant of the original owner, William Allen Lane, still lives at Lane plantation. Restoration was completed in 1969.

Laurel Hill (Felicianas)

Located in the northern section of West Feliciana Parish near the Mississippi state line, Laurel Hill manor is the expansion of a smaller house, which is the eastern half of the present home. A lower gallery runs across the entire front with ten square colonnettes. The home follows the style called Carolina I, which is quite different from most early Louisiana plantation homes. The first part was built about 1830; it was added to in 1873.

Laurel Hill was originally owned by Edward McGhee, the man who built the West Feliciana railway (the tracks still run across the property). In 1955 Mr. and Mrs. F. P. Farrar acquired Laurel Hill; when their house at Woodlawn burned in 1962 they moved to Laurel Hill. It is now a private cattle ranch.

Laurel Valley (Lafourche)

Laurel Valley plantation, built in 1834, exhibits the stuctures that made up the once thriving and productive community of a vast sugar plantation. Over seventy buildings, including cabins for skilled workers and unskilled workers, barns, a blacksmith shop, a school, servant's quarters, and the "big house," remain today as evidence of a prosperous domain. Sections of the large sugar house still remain even after extensive damages that occurred during Hurricane Betsy in 1965.

Laurel Valley began as a small farm owned by Etienne Boudreaux on property he obtained by a Spanish land grant in 1775. In 1834 the Boudreaux heirs sold the property to Joseph Tucker. Tucker purchased additional property and established the existing complex of facilities located nearly two miles from Bayou Lafourche.

Reportedly the old cabins, barns, and other outbuildings will undergo restoration to preserve them, and will eventually be opened to the public as a large-scale early Louisiana plantation museum. (Buildings along Bayou Lafourche at the junction of Parish Road 33 are a part of this plantation and will also be renovated for public admission.)

The main house, barely visible from the road through a grove of trees, is in good condition and still serves as the main manor for the present-day three-thousand-acre sugar plantation. The plantation is managed today by heirs of J. Wilson Lepine, Sr., who purchased Laurel Valley in 1893. (Side roads off the parish road are private and should not be entered without permission.)

This beautiful manor, built in 1850 facing the river at Arabi, no longer stands. Neglected and in disrepair for many years, it finally burned to the ground in the mid-1980s.

J DeHart

Lebeau

Les Memoirs (Attakapas)

An attractive early Louisiana home, Les Memoirs was built in 1836 by Aldelard de Rouselle. The lower section was originally open, exposing large square brick pillars, but was later enclosed to form the rooms of the lower floor.

Les Memoirs was restored to its present beauty by Mr. and Mrs. Merkel Stuckey. The beautiful tranquil grounds offer an appropriate setting for the St. Martin Parish plantation treasure.

L'Hermitage (Old River Road)

L'Hermitage is a splendid old mansion on the east bank of the river. It was built in 1812 by Marius Pons Bringier for his son Michel, who married the fourteen-year-old niece of the Abbe du Bourge of St. Louis Cathedral of New Orleans. The Bringiers were among the Creoles who fought with Jackson at the Battle of New Orleans. In Jackson's honor, the mansion was named after his Tennessee home.

The first crops of the plantation were indigo and tobacco; later the vast acreage was converted to sugar cultivation. The home and plantation were sold to Duncan Kenner in 1880. L'Hermitage is a private cattle ranch today.

Linwood (Felicianas)

A Louisiana classic home built by Gen. Albert G. Carter, Linwood dates to the early nineteenth century. Linwood played a very prominent role in the development of the Felicianas. It is now the manor of a large cattle ranch and is a private home.

jDeHart

Lloyd Hall

Live Oaks (False River)

Live Oaks was built at Rosedale in 1828 by Charles Dickinson. (Dickinson's father was killed years before in a duel of honor with Gen. Andrew Jackson.) After Charles died at the early age of forty, his wife Anna Maria Dickinson assumed the task of operating the plantation. She did this with a calm dignified strength and with as much skill as any man on a very successful basis for the next forty years.

The residence is two and one-half stories tall. It is situated in a grove of superb live oaks, one of which is nearly thirty feet in circumference. Within the grove is a slave church built of sunbaked bricks.

Lloyd Hall (Red River)

Located in southern Rapides Parish near Bayou Boeuf, Lloyd Hall is on the dividing line between the cotton and sugar industries of the state. Featuring unusual architecture, Lloyd Hall is in a fine state of preservation. It was built in 1816 supposedly for Lloyd's of London.

Today the old home houses a museum with old pieces and antiques. It is open to visitors by appointment.

Macland (Attakapas)

Macland was erected in the 1840s by Dr. Archibald Webb. It is located near Washington; the adjacent Bayou Courtableu and Bayou Boeuf furnished early transportation for crops from the four-thousand-acre plantation. The massive round columns, impressive stairs, gallery, grillwork, and the carriageway that runs through the center of the house have fallen into a state of disrepair.

jDeHart

Madewood

Madewood (Lafourche)

A well-preserved and imposing antebellum mansion, Madewood was built in 1840 for Col. Thomas Pugh of North Carolina. The house follows the Greek Revival tradition brought into Louisiana by American planters. It was designed and built by Henry Howard, the renowned American architect who built Nottaway. With complementing Greek Revival wings on each side, it boasts an outdoor kitchen of brick, other outbuildings, and an old family cemetery on the grounds to the rear.

Located on Bayou Lafourche near Napoleonville, Madewood is open to the public for an entrance fee.

Magnolia (Delta)

Two sea captains established Magnolia in 1780. George Bradish and William M. Johnson both married, lived together on Magnolia, and managed for some time the large sugar plantation and mill. In 1873 Henry Clay Warmoth, a former governor of Louisiana, bought the place and built the sixty-mile Buras to New Orleans railroad, according to some reports because his wife disliked steamboats and found horse and buggy travel too tiresome.

Magnolia is in a state of ruins, but the old original Magnolia store still exists directly across the road near the railroad tracks.

Magnolia (Red River)

The first home at Magnolia was burned by Union soldiers; shortly after the war, in 1868, the owner, Mathew Hertzog, rebuilt on the same foundation. Cotton was the site's prime crop, but Magnolia was also one of the few plantations that raised mules to sell at market.

Hertzog was an owner and breeder of fine race horses. On the grounds beneath a magnificent grove of magnolias and oaks the famous race horse Flying Dutchman is buried.

A row of brick slave quarters still exists at Magnolia in addition to the main house.

Magnolia Mound (Old River Road)

John Joyce of Fort Mobile, a merchant, contractor, and land developer, bought the thousand-acre plantation at Magnolia Mound and built the original mud and moss structure in 1791 to house his overseer. Joyce remained at Fort Mobile while the overseer managed the plantation. After Joyce died in 1798, his widow married Armand Allard Duplantier from Baton Rouge. They moved to Magnolia Mound and made extensive alterations to the original house.

The home was rescued from destruction in 1960 by the Foundation for Historical Louisiana. Guided tours are offered to the visitor for a small fee.

Magnolia Ridge (Attakapas)

Magnolia Ridge was built by slave labor in 1830 under the direction of Judge John Moore. After eleven years at Magnolia Ridge, Judge Moore married the widow of David Weeks, the builder of the Shadows in New Iberia. Both Confederate and Union forces used Magnolia Ridge as a head-quarters during the Civil War.

In 1938 the home was in a very sad state of deterioration when George Wallace rescued it and restored it to its original status. The present owners have cared very well for it and have opened it to visitors by appointment for a fee.

Malarcher (Old River Road)

The original mansion at Malarcher was built in the late 1700s by Le Chevalier Louis Malarcher, a political refugee of the French Revolution who became an influential citizen of St. James Parish. That home burned in 1890. The existing Creole cottage was built in 1891 by Willie Malarcher, the grandson of Louis Malarcher.

The home and plantation property were purchased in 1978 by Convent Chemical, a corporation jointly owned by B. F. Goodrich and the Bechtel Corporation. Beautifully restored by the company, the home will be used solely for business-related social activities and meetings for the company management. It is historically identified by a marker.

Mary (Delta)

Mary's West Indies architecture reflects the purest of Louisiana colonial plantation home designs. Built in the late 1700s, this hipped-roof historic gem is completely surrounded by two levels of cool spacious galleries. The first floor is constructed of brick with large columns; the second floor of virgin cypress with smaller colonnettes and simple balustrades. The entire home and grounds reflect an inviting, comfortable atmosphere.

Melrose (Red River)

Adjacent to Cane River Lake, Melrose was built by a son of Marie Coin Coin, a freed slave of Louis de St. Denis, the founder of Natchitoches. The home was first recorded as being owned by her mulatto son Louis Metoyer, whose descendants still live along the Cane River Lake.

The first house, which was known as Yucca, was built in 1796. Still standing, Yucca was used for many years as a guest house for the many well-known visiting artists and writers that began to come after John and Caminie Henry purchased Melrose in 1840. The present main house, constructed in 1833, is of original Louisiana colonial design with wings added to each side after 1840. Also on the grounds is the unusually designed African House, which was once used as a jail for unruly slaves.

One of the oldest and most famous of Louisiana plantations, Melrose is open to the public for a fee.

Mound (False River)

Mound was built on an ancient Indian mound by Austin Woolfolk in 1840. During the construction many Indian artifacts were uncovered along with an earthenware pot containing the skeletons of two babies.

Many stories have been told of the unusual kindness to slaves of the owners, Mr. and Mrs. Woolfolk. Still standing in the rear of the home is a brick slave laundry from the early days.

Nottoway

Mount Hope (Old River Road)

Erected by a German planter named Joseph Sharp in 1817, Mount Hope is one of the simpler homes of Louisiana plantations. It nevertheless boasts an unusual roof line and spacious gallery. The interior is splendidly furnished with early furniture and numerous other pieces of interest to the visitor. Open by appointment—check with the Baton Rouge tourist information center.

Mouton (Attakapas)

Framed with pines and magnolias, Mouton can be seen on North Sterling in Lafayette. It was built in 1848 by Charles Mouton, who later served as lieutenant governor of Louisiana.

The house was at one time occupied by Dr. Sterling Mudd, the nephew of Dr. Samuel Mudd. Samuel Mudd was the man sentenced to prison for setting the broken leg of Lincoln's assassin John Wilkes Booth.

Myrtles (Felicianas)

The Myrtles features unusual architecture dating from 1830. Gen. David Bradford obtained a Spanish land grant of 650 acres when he came to Louisiana; on that land he built the Myrtles. It is lavishly furnished with many antiques and supposedly has ghosts of early Louisiana. Open to the public; admission for a fee.

Nottoway (Old River Road)

Nottoway is the largest existing plantation home in Louisiana. It has been beautifully restored by Mr. Arlin K. Dease, who has made the home his residence. The magnificent sixty-four room mansion was a showplace of the South and was surpassed in size only by its rival Belle Grove, which has since been taken by fire.

Nottoway was built in 1857 for John Hampden Randolph by one of the greatest architects of the nineteenth century, Henry Howard. It was named after a county in Randolph's home state of Virginia.

Oak Alley

The magnificent White Ballroom was on many occasions the center of fabulous balls, weddings, and social gatherings. The home is of an unusual architecture not normally seen in Louisiana plantation homes. There are several supporting buildings, including separate quarters for the Randolph daughters and a *garçonnière* for the sons.

Nottaway is open to the public for a fee. It has an accommodating restaurant and a gift shop.

Oak Alley (Old River Road)

Oak Alley is one of the best known of Louisiana plantations. The old classic home stands at the end of a beautiful double row of twenty-eight ancient oaks. On the west bank River Road above Vacherie, it is in a setting appropriate to Louisiana's great past.

Built in 1836 by Jacques T. Roman, Oak Alley has been admired by travelers along the old river for almost a century and a half. Many of the early riverboats, loaded with passengers from the north, would slow down for better view of the grand manor. Its original name was Bon Sejour— pleasant visit—but the travelers along the river gave it the name of Oak Alley.

Even today the National Historic Landmark captures the hearts of thousands of travelers. The mansion and grounds are open to the public for a fee. There is a restaurant for visitors.

Oakland (Dixie-Overland)

Located near Houghton in north Louisiana, Oakland stands in a grove of oaks atop a hill. It is an old cotton plantation with many haunting stories, perhaps the most familiar of which of ghostly horsemen sweeping over the hills at night on mounts with thunderous hooves.

The builder, Dr. Abel Skannel, is said to have had his own coffin made years before his death. According to stories, he often would entertain friends by laying in the coffin, demonstrating to them how well he fit. The Negroes became very superstitious about him and created many haunting stories concerning the old home. Dr. Skannel prospered in his medical practice and his plantations. When he died he owned a total of five plantations: Oakland, Sligo, Cave Bend, Bluff, and Chalk Level. Sligo was his largest.

Oaklawn

Oakland (Red River)

On Cane River Lake below Natchitoches, Oakland, a large Louisiana colonial home, was erected in 1818 by Pierre Emmanuel Prudhomme, a very successful planter of cotton, tobacco, and indigo. Prudhomme was the first planter to grow cotton on a commercial basis in Louisiana.

The house was constructed by slave labor using only the finest of heart cypress, joined together without a single nail. The walls are of bousillage construction with handmade bricks at the bottom. A ten-foot-wide gallery completely encircles the house, at the front and rear stretching eighty feet and at the sides seventy feet.

Oakland was used in the filming of the movie "The Horse Soldiers," starring John Wayne. It is open to the visitor by appointment only.

Oaklawn (Techeland)

On Bayou Teche near Franklin is Oaklawn, a beautifully restored mansion surrounded by a huge grove of old live oaks. With over 250 live oaks, the grove is said to be the largest in America.

The house was built in 1827 by Alexander Porter, an Irish immigrant. Porter, one of Louisiana's early statesmen, helped write the state constitution. He served in the state legislature, the Louisiana Supreme Court, and the U.S. Senate. He built the mansion at Oaklawn after his retirement from public life.

Oaklawn has an interesting and colorful history as a huge sugar plantation. Across Bayou Teche from the manor is the sugar house (on Highway 87), which still processes sugar from the vast lands of the Oaklawn cane fields, extending for miles in all directions.

Oaklawn was restored by Capt. C. A. Barbout in 1926; more recent work has been supervised by Mr. and Mrs. George B. Thomson. The beautifully shaded grounds and richly furnished mansion are well worth visiting, and are open to the public for a fee.

Oakley (Felicianas)

Located in Audubon Memorial State Park in the Felicianas is Oakley, built in 1810 on a Spanish land grant obtained in 1770. The builders, James and Lucy Perrie, hired a tutor for their young daughter Eliza from New Orleans to teach drawing, music, dancing, and other fine pursuits. He was paid a sum of sixty dollars a month for this work, including room and board. That tutor was John James Audubon, who later became famous for his birds of America paintings. Oakley is now part of a state park named in honor of Audubon and is open to the public for a nominal fee.

Oakwold/Wright (Red River)

The beautiful old home known as Oakwold or Wright was built in 1835 by S. M. Perkins, an ancestor of the Wright family, which later gave the house its name. Entertainment at Wright plantation was always in a fashion of elegance. On at least two occasions it is said that Sam Houston was a guest at the social events at Wright.

Oakwold/Wright is a private residence today on shaded grounds at the outskirts of the peaceful little town of Evergreen.

Ormond (New Orleans)

On the east bank River Road a short distance north of Destrehan is Ormond, a good example of an early Louisiana colonial plantation home. It resembles Melrose very much with the two added wings.

Ormond's builder and master, Pierre Trepagnier, a Frenchman of nobility, honorably distinguished himself in the defense of Louisiana under Gov. Bernardo de Galvez when his forces successfully defeated the British in 1789. Trepagnier's noble efforts were rewarded with vast plantation lands that extended from the Mississippi River to Lake Pontchartrain. On this newly acquired land he built his home in the late 1790s.

One morning in 1798 Trepagnier had just sat down to breakfast with his family of eight when a stranger rode up on horseback. Trepagnier went out to greet the stranger and conversed with him briefly. He returned to tell his family that he was going on an errand and promised to return in a short while. He left on horseback with the caller. Trepagnier was never again seen or heard from after his strange departure; the mystery of his disappearance was never solved.

A year later, after anxious months and dreadful nights waiting for Trepagnier's return, his bereaved wife, exhausted with grief, abandoned Ormond. She took her children with her, leaving the plantation and mansion empty and neglected.

In 1812 Richard Butler, a former officer in the Continental Army under the Marquis de Lafayette, bought Ormond to become a gentleman planter. He brought his new bride, the daughter of the Spanish governor of Natchez, to make their home in the riverside manor. The Butlers gave the name "Ormond" to the mansion, naming it after one of Richard's relatives, the Earl of Ormond.

His fortunes grew in a few short years and the marriage was a happy one even without children. But one disaster was followed by another: bad crops, floods, storms, and finally a yellow fever epidemic claimed the lives of both Butler and his young wife.

A friend of Richard Butler, a sea captain by the name of Samuel McCutchon, purchased Ormond from Butler's relatives. He then married one of Butler's sisters, became a planter, and filled the house with a growing family—nine children in rapid succession. The McCutchons added the twin-wing *garçonnière* and enjoyed great success with the plantation.

Ormond

The old house later fell into decay as a result of natural accidents and the ravages of the Civil War, but in the late 1890s a family by the name of LaPlace restored Ormond to peaceful prosperity. However, tragedy struck again in almost the same fashion as once before, for one night while seated peacefully at the supper table with his family, Mr. LaPlace was visited by a strange caller. He left the house on what his wife said was a routine short errand, but he never returned to his supper. On the following morning his bullet-riddled body was found hanging from the great oak that shades the front yard. As with the earlier disappearance, the mysterious crime was never solved. Again, Ormond was abandoned to nature's destructive elements; once again, it seemed that Ormond was cursed with sadness and neglect.

The home has recently been restored in a new challenge to the uncanny forces that seem to surround the site. A private home, Ormond has been treated to new paint and new furnishings.

Palo Alto (Lafourche)

Palo Alto is a peaceful, well-preserved old home that nestles beneath huge ancient live oaks in the Lafourche country. Built in 1850, it is portrayed in a painting made by the famous Louisiana artist, Adrienne Persac. Persac stayed at the plantation while making the painting shortly after Palo Alto was built.

There are many beautiful antiques at Palo Alto, including some that once belonged to famous Louisianians. One such treasure is an original of the famed Norman's Chart of the Mississippi, which shows all of the early plantations from New Orleans to Natchez. The map was actually created by Adrienne Persac, but it was published by a man named Norman in 1858.

Parlange (False River)

Parlange is one of the most popular homes of Louisiana colonial design. It was built in 1750 on a land grant issued to the Marquis Vincent de Ternant. The intriguing home has housed six generations of the builder's family descendants, who have lived in Parlange refusing to allow it to fall victim to the ravages that have destroyed so many Louisiana treasures. It was originally an indigo plantation, but the crop was later changed to sugar.

It is reported that three separate treasures of gold and silver were buried on the grounds shortly before Union soldiers overran the place during the Civil War. However, only two have been recovered.

Galleries completely encircle the house, which is listed as a National Landmark by the U.S. Department of the Interior, and twin brick *pigeonniers* flank each side. Parlange is located on False River a few miles below New Roads and is open to the public for a nominal fee.

Poiret/Means (Attakapas)

French aristocrat Chevalier Florentin Poiret built the early Louisiana colonial home bearing his name that now stands elegantly among magnolias and live oaks. A private residence, Poiret can be viewed from Plaisance Road, near Opelousas.

Riche (False River)

A Louisiana colonial plantation home located in Pointe Coupee Parish, Riche has been preserved since its construction in 1825. Riche was an early cotton and sugar plantation and was owned and operated by Fannie Riche, a Negro. During the Civil War the home was used as a hospital for soldiers. The home is now privately owned by Mr. and Mrs. Gerald P. Schexnayder, Sr.

Rienzi (Lafourche)

Records indicate that this home was built in 1796 as a refuge for Queen Marie Louisa of Spain in the event of Spain's defeat in the Napoleonic Wars. It is of modified Louisiana colonial design, with imposing stairways in the front (similar to those of Evergreen). Rienzi was later owned by Henry Schulyer Thibodaux, the founder of the town of Thibodaux. The plantation ownership eventually passed into the hands of the J. B. Levert Land Company.

Ringrose (Attakapas)

Ringrose was built in 1770 by Michel Prudhomme, a pioneer in the Attakapas country. Its design is characteristic of early Louisiana colonial, with cypress above brick, large brick columns on the lower floor, and hand-hewn cypress colonnettes on the spacious galleries of the second level. Plans are being made to renovate the house and open it to the public.

Rosebank (Felicianas)

Spanish Commander Don Juan O'Conner built Rosebank in the 1790s. The unusual Louisiana colonial home was at one time called the Spanish Inn. Cypress colonnettes and balustrades have been replaced with iron posts and railings.

Rosedown (Felicianas)

An 1835 Louisiana classic home on magnificent grounds, Rosedown is a visitor's showpiece. Elegant furnishings, formal gardens, and superb live oaks bring the old home back into the era of the great plantations, when sugar and cotton reigned supreme throughout the South.

The home was built by Daniel Turnbull for his bride, who was of the Barrows family. The Barrowses owned the grand Greenwood plantation, which was located northeast of St. Francisville near the Mississippi River.

Rosedown is open to the public for a fee.

Rosedown

DeHart

Rosella (Lafourche)

Rosella was built in 1814 by Jean Baptiste Thibodaux. His widow Natalie then married Everiste Lepines; both were buried on the Rosella grounds after their deaths. Rosella was spared from destruction in the Civil War because of a personal order given by General Sherman himself, a close friend of the owner.

There are many plantation and household items still remaining that are original to the plantation. A tranquil beauty of Bayou Lafourche, Rosella is private but can be visited at Spring Fiesta time and during the Sauce Piquante Festival in Raceland each October.

Roseneath (Dixie-Overland)

William Bundy Means, the son of a prominent South Carolina family, came to the fertile lands of north Louisiana and built his home between the Texas border and the Red River on the first large plantation in the newly acquired Indian lands. Roseneath took three years for him to build with careful slave labor, and was completed in 1840. One hundred forty years later, the house has settled only one-half inch. The old home is not characteristic of Louisiana design, although the galleries are deep and halls quite wide.

Roubieu/Reform (Red River)

Roubieu/Reform was built in 1808 by Francois Roubieu, a Frenchman. Roubieu was one of the first white men to build on Isle Brevelle, for until then there were only free people of color living in the vicinity. The handsome and sturdy old home has been well preserved and has recently been renovated to its original beauty.

Sacred Heart Academy and St. Charles College
(Attakapas)

In the peaceful little town of Grand Coteau, located between Lafayette and Opelousas in the very center of Acadiana, is the Academy of the Sacred Heart, a college for young girls. The academy is the second-oldest school of higher learning in Louisiana, preceded only by the Ursuline Convent in New Orleans. Of 212 Sacred Heart schools around the world, this academy is the oldest in continuous operation.

A fine institution of cultural and academic values, Sacred Heart was founded by the Sisters of the Sacred Heart to educate the daughters of the planters of the Attakapas and Techeland region. In spite of yellow fever epidemics, devastating fires, ravaging floods, famines, and the disasters of the Civil War, it has served in this capacity continuously for over 160 years.

Mother Eugenie Aude, a former member of the Napoleonic court in France, and Sister Mary Layton came from St. Louis by boat down the Mississippi and into the bayous, and then overland by oxcart and on horseback until they reached Grand Coteau. On October 5, 1821, with only eight students enrolled, Mother Eugenie and Sister Mary opened the academy in a two-story building on land donated by Charles Smith.

The original two-story structure was destroyed by fire in 1922. The present three-and-one-half story brick main building is another old building, built in 1830. At the same time as the construction of the present main building the school's famous gardens, patterned from the French garden of Bishop Bossuet, were laid out. Other adjacent buildings were later added.

In 1838 St. Charles College, a Jesuit school for young men, was built near Sacred Heart with bricks donated by the academy. The school was destroyed by fire in 1908, but it was immediately rebuilt, in the form that stands today. Other buildings were later added and the present-day complex of the two schools and St. Charles Church, together with supporting buildings, share the serene shades of beautiful pines and oaks situated on one thousand acres of land in northeastern Grand Coteau. The famous avenue of 140-year-old live oaks was planted by the Reverend Nicholas Point, the Jesuit priest who was the first rector at St. Charles. The son of Union General William T. Sherman, a priest at St. Charles, is buried in the cemetery.

In 1866 a young postulant at the academy named Mary Wilson became critically ill. The entire community prayed in novena to St. John Berchmans for nine days and the girl's miraculous recovery was attributed to his help. The infirmary was converted into a chapel in honor of the miracle. Sister Mary Wilson's grave is in the old academy cemetery nearby.

The Academy of the Sacred Heart is now on the National Register of Historic Places. St. Charles College is a seminary to educate young men for Jesuit priesthood. It was changed from an open college to a seminary in 1922 and remains serving in that capacity today.

St. Emma (Lafourche)

St. Emma, an early Louisiana colonial, was built by Charles Koch of Belgium the same year that he built Belle Alliance a few miles away across Bayou Lafourche.

St. John (Attakapas)

Near St. Martinville, St. John was erected in 1828 by General Alexandre Etienne DeCluet, a descendant of the army commander and lieutenant governor of the Attakapas. The mansion is a very imposing structure on the east bank of Bayou Teche north of St. Martinville. With its grand manor, sugar mill, and many supporting buildings, St. John is still an operating sugar plantation.

St. Joseph (Old River Road)

On the old River Road near Vacherie, St. Joseph was built by Dr. Cazimis Merciq in 1820. It was later purchased by the wealthy Valcour Aime as a wedding present for his daughter Josephine.

Another of the early Louisiana colonial homes that were popular in the late 1700s and early 1800s, St. Joseph features a hipped and dormered roof, wide galleries, and brick and cypress construction with bousillage walls.

J DeHart

San Francisco _____

St. Louis (Old River Road)

Originally called Home plantation, St. Louis was established by Capt. Joseph Erwin of Tennessee, who made and lost fortunes as a sugar planter. The first home was destroyed by the Mississippi. The existing mansion was built in 1858 by Edward J. Gay of St. Louis, who named the place for the city that he loved.

St. Louis is a tall structure of Louisiana classic design with Corinthian columns and ornate Creole style iron grillwork on the galleries. The home is the center of an operating sugar plantation.

San Francisco (Old River Road)

On the old east bank River Road near Reserve is an interesting and imposing structure, perhaps the most ornate of all Louisiana plantation homes. San Francisco's design has been dubbed "Steamboat Gothic," because it reflects architecture of the old river steamboats. The mansion is the focus in the 1952 novel *Steamboat Gothic* by the famous authoress Frances Parkinson Keyes.

Built in 1850 by Edmond Marmillion, who died before the house was completed, San Francisco is in no way related to the city of that name in California. The name came into existence as a result of continued mispronunciation of its true name, Sans Frusquin.

The beautiful frescoes on the ceilings were painted by Dominique Canova, the nephew of Napoleon's favorite sculptor. Those frescoes were recently restored as extensive renovations were made throughout the house and grounds by Koch and Wilson Architects of New Orleans. The beautiful old home features gingerbread ironwork, unusual windows, belvederes, double stairs, huge cisterns, frilly decor, and lavish furnishings. It is open to the general public for a fee.

Shadows on the Teche (Techeland)

The Shadows on the Teche is perhaps written about more than any other plantation home of old Louisiana. Such writers as Harnett Kane and Lyle Saxon have immortalized the home in their stories of plantation life. Together with Melrose and Oak Alley, it is one of the most popular homes for visitors.

The Shadows was built on the banks of Bayou Teche in New Iberia by David Weeks in 1831. Weeks, a prominent sugar and cotton planter, established the home and fields on land granted to

Shadows·on·the·Teche

j DeHart

Southdown

his father in 1792. With his wife Mary Clara Conrad, David Weeks ruled an empire including Parc Perdu, Cypremort, Shell Mound, Alice, Acadie, Richohoc, Town Farm, and Week's Island. He built the Shadows on Town Farm plantation, which was located near a settlement that later became the city of New Iberia.

Situated in a very serene setting amid moss-shrouded oaks and tropical plants, the rose-colored brick mansion dominates its surroundings. A beautiful Louisiana classic home in the center of the "Queen City of the Teche," the Shadows has been placed in the care of the National Trust for Historic Preservation. The home contains treasures, paintings, furnishings, and silver from four generations of the family. Declared a National Historic Landmark by the Department of the Interior, Shadows on the Teche is open daily to the public for a nominal fee.

Southdown (Terrebonne)

Southdown, an English manor erected in 1858, appears very different in design from the homes normally seen on plantations of the Old South. Erected by Stephen Minor, the former secretary of Spanish Governor Gayoso de Lemos who later became governor of Natchez, Southdown was originally a one-story building with one-foot-thick walls. The second story level and turrets were added in 1898. To the rear is a two-story brick building that originally housed a kitchen, dairy, laundry, and servant quarters.

Southdown acreage was first planted with indigo, but shortly afterwards sugar was substituted. It rapidly became one of Louisiana's largest sugarcane plantations.

On plantations, grass and weeds were always a problem. To save the labor of constant hoeing by slaves, animals were used extensively to keep the grass cut. On sugar plantations many of the planters used sheep for this purpose, since the animals did not eat sugarcane but did a magnificent job of keeping all weeds from growing around the plants.

Southdown was one of the many plantations that used sheep for this purpose. The plantation was in fact named after a famous English breed of sheep that the Minors introduced to Louisiana. The plantation business was managed by Stephen Minor's son William J. Minor, who was very well known for his high living and lavish entertaining.

Southdown has now been converted into the Terrebonne Museum and is open to the public.

Starvation Point (Attakapas)

Built in 1790 by Luc Collins, a native of Hampshire County, Virginia, Starvation Point home overlooks the junction of three bayous, Cocodrie, Boeuf, and Courtableu. Steamboats from the lower Teche would come as far up as this junction. The two-story classic home, now private, emits a ghostly atmosphere of loneliness.

Sunnybrook (New Orleans)

Listed on the National Register of Historic Places, Sunnybrook is a captivating old home situated on peacefully shaded grounds in the center of one of Louisiana's most beautiful areas, St. Tammany Parish. Sunnybrook is a good example of a typical Louisiana plantation home of the nineteenth century. Built in 1880 by a German immigrant named Fritz Buchin, it is now owned by Mr. and Mrs. William Gibert, who have carefully preserved it.

Susie (Techeland)

A Louisiana classic home with matching front and rear galleries and pillars, Susie was built in 1852 by Roal Harris. It was rescued from deterioration in 1970 by the Sutters, who also own Bocage, located a short distance away across the highway.

Tallyho (Old River Road)

The home now known as Tallyho was once the overseer's home for the great Tallyho plantation, located above White Castle. When the main house burned to the ground the family took over the present Tallyho home.

Tezcuco (Old River Road)

Tezcuco takes its name from the Aztec village in Mexico whose name means "resting place." It was built in 1855 for Benjamin Turead, a veteran of the Mexican War.

The large Creole cottage has front bedrooms that are twenty-five feet square and ceilings fifteen feet high. It is elegantly furnished with many fine antiques. Beautiful ancient live oaks and magnolias surround the old home.

Theriot/Khi Oaks (Lafourche)

Alexander Theriot built Khi Oaks in 1890. At the time it was said to be the most modern home of the Bayou Lafourche area, perhaps because it featured indoor plumbing. After having been abandoned for many years, it was restored to its original beauty in 1972 by the W. W. Wilson family. The old home is surrounded by moss-shrouded oaks that are members of the Live Oak Society of America.

Trinity (False River)

Trinity was built in 1839 by Dr. George Washington Campbell of New Orleans. It stands on an ancient Indian mound at the end of an avenue of fine oaks.

Dr. Campbell and his family lived in New Orleans at the corner of St. Charles Avenue and Julia Street, where he maintained a very successful medical practice. Early in May of 1863, during the fall of New Orleans to Farragut's fleet, Dr. Campbell was away at Trinity plantation attending to business affairs while his family remained in New Orleans. Late one night, Mrs. Campbell and her five children were rudely awakened by Gen. Benjamin Butler and his troops. They were instructed to clothe themselves and vacate the house immediately for Butler's use. They were allowed to take only the clothing that they wore. They joined Dr. Campbell at Trinity, never again to return to New Orleans.

On the grounds are ruins of supporting structures, including an old slave hospital. The original plantation church, now restored, has been moved into nearby Rosedale, across the bayou, and can be visited by the public.

Variety (Old River Road)

The old Variety home, constructed in 1856, was destroyed by fire in 1973. Also destroyed in that fire along with the elegant mansion were many fine antique furnishings, and some of the ancient live oaks, sweet olives, and magnolias that graced the grounds of the plantation.

To help ease the sorrow and heartache of the losses for his mother, owner John W. Wilbert, Jr., replaced old Variety home with another plantation home that bears an amazing resemblance to the first with the major exception that it is smaller in size. Mr. Wilbert had the present home and guest house, once a part of Homestead plantation at Turnerville, moved twelve miles to this location. The present Variety home, built in 1828, is older by nearly thirty years than old Variety. Mr. Wilbert also planted a seventy-five-year-old, five-thousand-pound sweet olive tree on the grounds as a monument to the ancient trees that once lived on the plantation. A two-hundred-year-old live oak at the rear of the home still exists, but has sustained scars from the devastating fire.

Villa de la Vergne (New Orleans)

A quaint plantation manor with an unusual bell tower, Villa de la Vergne snuggles peacefully in a beautiful setting on the east bank of the Bogue Falaya River near the town of Covington, on the north side of Lake Pontchartrain. Built by slave labor in 1820, the French villa features vine-covered grounds and the tranquil atmosphere prevalent throughout St. Tammany Parish.

Weynoke/Beechwood (Felicianas)

The original house at Weynoke was a one-story log cabin. In about 1809 Capt. Robert Percy built the present two-story home, said to be the Beech Woods house in which Lucy Bakewell Audubon conducted school in 1820. Perhaps in the nearby woods her husband, John James Audubon, found subjects for many of his famous bird paintings, including "Wild Turkey."

Villa de la Vergne

J DeHart

E. D. White (Lafourche)

E. D. White plantation home was the birthplace and home of Chief Justice Edward Douglass White, Jr., of the U.S. Supreme Court. The home and sixteen hundred acres of farm land were purchased by the U.S. Government but were later turned over to the State of Louisiana for use as a state museum and park. The home and grounds are now open to the public.

The home exemplifies the typical Louisiana Creole raised cottage. On the grounds is the famous Blanchard Oak, which is registered with the Louisiana Live Oak Society as a member. The society has a human chairman and an attorney whose duty is to protect the vast number of oaks in Louisiana. Membership in the society, which is composed entirely of trees, is granted by age, girth, and acorn production.

Whitehall (False River)

Recently rescued from deterioration, Whitehall is in the northwestern tip of Pointe Coupee Parish on the east bank of the Atchafalaya River. It was built by Gen. B. B. Simmes, the founder of Simmesport, the town across the river from Whitehall on the west bank of the Atchafalaya. The two-story classic structure boasts wide upper and lower galleries fronted by six square columns and balustrades. A large lower floor bay window protrudes on the north side.

Whitney (Old River Road)

Whitney, located below Vacherie, was built in the early 1800s. A typical early Louisiana plantation dwelling, it is two story with brick lower floor and cypress above. All hardware of the old home was handmade by slaves.

The artistic work on the parlor ceiling and doors was done in gratitude by a thankful artist. He had been commissioned by the Catholic church nearby to do art work, but became very ill. Mrs. Haydel, then the mistress of Whitney, cared for him and nursed him back to health.

Winter Quarters (Tensas)

With an unusual design for a Louisiana plantation home, Winter Quarters resembles the main house of an old west ranch. Built in 1852 by Haller Nutt, whose wife was related to the famous Jim Bowie, there is a local tradition that claims that the home was used by General Grant as his winter quarters during the Vicksburg operations. Although Grant possibly could have spent a few days there in 1863, legal records bear out that the name "Winter Quarters" was used long before the Civil War. The plantation could have obtained the name from the fact that the owners left during the summer months, moving to Natchez to avoid the marshy fever-ridden area during the hot season, returning only during the winter months.

Woodland (Delta)

William M. Johnson, an original co-owner of Magnolia plantation downriver, built this home in the early 1800s. Johnson and George Bradish owned Magnolia together and lived there with their wives until trouble between the women caused a split. Johnson sold his share of Magnolia to Bradish and established a new home for himself, his wife, and his son, christened Bradish Johnson (he was born before the ill feelings between the two couples).

Woodland is a large raised Creole cottage that is in a very deteriorated condition. It shows evidence of having been a very handsome place, but has been sadly neglected over the past years. It faces the Mississippi River; the rear can be seen by the traveler on the highway.

Zachary Taylor (False River)

Zachary Taylor is a large Louisiana classic home that has been very carefully restored. The plantation is said to have been operated by the brother of Zachary Taylor; one of the former president's daughters also spent most of her time at this old place. Taylor himself lived on another of his plantations, located across the river from Baton Rouge, prior to his presidency.

INDEX
Existing Homes and Institutions

Existing Homes and Institutions _____